S0-BJR-109

TREATISE ON BASIC PHILOSOPHY

Volume 2

SEMANTICS II: INTERPRETATION AND TRUTH

TREATISE ON BASIC PHILOSOPHY

1

SEMANTICS I *Sense and Reference*

2

SEMANTICS II *Interpretation and Truth*

3

ONTOLOGY I *The Furniture of the World*

4

ONTOLOGY II *A World of Systems*

5

EPISTEMOLOGY I *The Strategy of Knowing*

6

EPISTEMOLOGY II *Philosophy of Science*

7

ETHICS *The Good and the Right*

MARIO BUNGE

Treatise on Basic Philosophy

VOLUME 2

Semantics II:

INTERPRETATION AND TRUTH

D. REIDEL PUBLISHING COMPANY

DORDRECHT-HOLLAND / BOSTON-U.S.A.

Library of Congress Catalog Card Number 74–83872

ISBN 90 277 0535 6

Published by D. Reidel Publishing Company,
P.O. Box 17, Dordrecht, Holland

Sold and distributed in the U.S.A., Canada, and Mexico
by D. Reidel Publishing Company, Inc.
306 Dartmouth Street, Boston,
Mass. 02116, U.S.A.

Printed in The Netherlands by D. Reidel, Dordrecht

GENERAL PREFACE TO THE *TREATISE*

This volume is part of a comprehensive *Treatise on Basic Philosophy*. The treatise encompasses what the author takes to be the nucleus of contemporary philosophy, namely semantics (theories of meaning and truth), epistemology (theories of knowledge), metaphysics (general theories of the world), and ethics (theories of value and of right action).

Social philosophy, political philosophy, legal philosophy, the philosophy of education, aesthetics, the philosophy of religion and other branches of philosophy have been excluded from the above *quadrivium* either because they have been absorbed by the sciences of man or because they may be regarded as applications of both fundamental philosophy and logic. Nor has logic been included in the *Treatise* although it is as much a part of philosophy as it is of mathematics. The reason for this exclusion is that logic has become a subject so technical that only mathematicians can hope to make original contributions to it. We have just borrowed whatever logic we use.

The philosophy expounded in the *Treatise* is systematic and, to some extent, also exact and scientific. That is, the philosophical theories formulated in these volumes are (*a*) formulated in certain exact (mathematical) languages and (*b*) hoped to be consistent with contemporary science.

Now a word of apology for attempting to build a system of basic philosophy. As we are supposed to live in the age of analysis, it may well be wondered whether there is any room left, except in the cemeteries of ideas, for philosophical syntheses. The author's opinion is that analysis, though necessary, is insufficient – except of course for destruction. The ultimate goal of theoretical research, be it in philosophy, science, or mathematics, is the construction of systems, i.e. theories. Moreover these theories should be articulated into systems rather than being disjoint, let alone mutually at odds.

Once we have got a system we may proceed to taking it apart. First the tree, then the sawdust. And having attained the sawdust stage we should move on to the next, namely the building of further systems. And this for three reasons: because the world itself is systemic, because no idea can become fully clear unless it is embedded in some system or other, and because sawdust philosophy is rather boring.

The author dedicates this work to his philosophy teacher

Kanenas T. Pota

in gratitude for his advice: "Do your own thing. Your reward will be doing it, your punishment having done it".

CONTENTS OF *SEMANTICS II*

PREFACE TO *SEMANTICS II*

This is the second and last part of our work on semantics. The first part, titled *Sense and Reference*, constitutes volume 1 of the *Treatise*.

What follows presupposes an understanding of the slippery notions of sense and reference. Any theories elucidating these concepts will do for the purpose of tackling the present volume. But, of course, only the theories expounded in Part I will articulate cogently with those we are about to discuss. Nevertheless, the gist of Part I can be summarized in a few words.

Philosophical semantics is about constructs, in particular predicates and propositions. Every such object has both a sense and a reference. The full sense of a construct is the collection of its logical relatives. This collection is made up of two parts: the purport, or set of implicants, and the import, or set of implicates. For example, the purport of a defined concept is the set of concepts entering in its definition, and its import is the collection of concepts hanging from it. As to the referents of a predicate, they are the individuals occurring in its domain of definition. And the reference class of a statement is the union of the reference classes of all the predicates occurring in the proposition. Some constructs, notably those occurring in ordinary knowledge and in scientific theories, have a factual sense and a factual reference. The theories of sense and reference put forth in Part I allow one to calculate both the sense (in particular the factual sense) and the reference class (in particular the factual reference class) of any predicate and any statement. By this token they can help us solve certain tricky semantical problems posed by some of the most important scientific theories, the sense and reference of which are often the object of spirited debates. So much for a résumé of Part I.

The present volume starts with the problem of interpretation. Interpretation is construed as the assignment of constructs (e.g. predicates) to symbols. It can be purely mathematical, as when the dummy x is interpreted as an arbitrary natural number, or also factual, as when such a number is interpreted as the population of a town. Now, as we saw a while ago, predicates and propositions have both a sense and a reference – and nothing else as far as meaning is concerned. These, then, are taken to be the meaning components. That is, the meaning of a construct is defined as the ordered couple constituted by its sense and its reference class. Once the meaning of a proposition has been established we can proceed to finding out its truth value – provided it has one. If the proposition happens to be factual, i.e. to have factual referents, then it may be only partially true – if true at all. Hence we must clarify the concept of partial truth of fact. This we do by building a theory that combines features of both the correspondence and the coherence theories of truth. The remaining semantical notions, notably those of extension, vagueness, and definite description, are made to depend on the concepts of meaning and truth and are therefore treated towards the end of this work. The last chapter explores the relations between philosophical semantics and other branches of scholarship, in particular logic and metaphysics.

This volume, like its predecessor, has been conceived with a definite goal, namely that of producing a system of philosophical semantics capable of shedding some light on our knowledge of fact, whether ordinary of scientific. We leave the semantics of natural languages to linguists, psycholinguists and sociolinguists, and the semantics of logic and mathematics (i.e. model theory) to logicians and mathematicians. Our central concern has been, in other words, to clarify and systematize the notions of meaning and truth as they occur in relation to factual knowledge. For this reason our semantics borders on our epistemology.

SPECIAL SYMBOLS

C	Set of constructs (concepts, propositions, or theories)
$\mathbb{C}$	Context
$\mathscr{C}$	Content (extralogical import)
$\mathscr{C}n$	Consequence
$\mathscr{D}$	Designation
Δ	Denotation
$\hat{=}$	Representation
$\mathscr{E}$	Extension
$\mathscr{I}$	Intension
$\mathscr{I}mp$	Import (downward sense)
L	Logic
$\mathscr{L}$	Language
$\mathscr{M}$	Meaning
Ω	Universe of objects (of any kind)
$\mathbb{P}$	Family of predicates
$\mathscr{P}ur$	Purport (upward sense)
$\mathscr{R}$	Reference
S	Set of Statements (propositions)
$\mathscr{S}$	Sense
$\mathscr{S}ig$	Signification
T	Theory (hypothetico-deductive system)
$\mathscr{V}$	Truth value function

CHAPTER 6

INTERPRETATION

All of the symbols in a scientific theory are interpreted. They are interpreted as designating certain mathematical concepts, some of which are in turn interpreted as representing certain aspects of the world. Such a double interpretation must be shown as completely and explicitly as possible if the signification of the symbolism is to emerge clearly. But what is an interpretation, in particular a factual one? This is the central theme of the present chapter.

1. Kinds of Interpretation

Anything, from sign to gait, can be interpreted if one knows how to. Thus farmers interpret cloud shapes, physicians bodily appearances, and charlatans dreams. In all three cases observed facts are correlated to hypothesized ones and the latter are assumed to explain the former. This kind of interpretation, bearing on natural signs, may be called *epistemic*: actually it is a mode of explanation. The concern of semantics is with another kind of interpretation, one bearing either on signs or on constructs. Henceforth we shall adopt this acceptation of 'interpretation', which may be called *semiotic*.

Semiotic interpretation may be construed either as bearing on signs or as bearing on constructs. The interpretation of signs is a task for designation rules, while the interpretation of constructs is performed by semantic assumptions. Example of sign interpretation: '&' *designates* conjunction. Example of construct interpretation: $F(a, b)$ *represents* the strength of the interaction between a and b.

Whether bearing on signs or on constructs, interpretation is required whenever that which is interpreted is not definite enough. Interpretation goes from the less to the more definite or specific. For example, from an ambiguous sign like 'S' to a definite generic construct such as "set", from the latter to a specific construct like "the set of pairs", or from here to a factual item like the collection of married couples, or to an empirical item such as the collection of married couples counted by the census bureau. We distinguish then four different kinds of interpretation relations, that are exhibited and exemplified in Table 6.1.

TABLE 6.1

Semiotic interpretations

Kind of interpretation		Relata	Example
1 Designation	$\mathscr{D}$	Symbol→Construct	Function letter→Function
2 Mathematical	μ	Generic construct→Specific construct	Function→sin
3 Factual	ϕ	Specific construct→Factual item	$\sin \omega t$→pendulum elongation
4 Pragmatic	π	Specific construct→Empirical item	$\sin \omega t$→measured value of pendulum elongation

The first kind of interpretation, i.e., designation, occurs in every conceptual system: without designation rules a symbolism does not symbolize. Thus a page of the *Journal of Mathematical Psychology* may be regarded as a system of conventional signs (words and mathematical symbols) together with a set of interpretation conventions – mostly tacit but nevertheless operative. In other words, a conceptual system may be regarded as an *interpreted language*, i.e., a symbolism together with a collection of designation rules. An uninterpreted language, i.e., a well constructed system of artificial signs with no designata, would be as idle and unintelligible as a scientific manuscript after a total nuclear holocaust. The very notion of a totally uninterpreted language makes no sense except for purposes of analysis.

The most basic of all conceptual systems are of course the logical calculi: they are the most abstract in the sense that they are the least interpreted. Logical calculi are prime examples of *abstract theories*, i.e. of theories containing predicates with no fixed interpretation – hence making room for a variety of interpretations. But they are all interpreted languages in the sense that they contain a designation rule for every sign type. Thus a predicate letter like '*P*' is interpreted as an arbitrary predicate or attribute. The interpretation is limited to designation: the calculus is uninterpreted solely in the sense that it involves none of the kinds 2 to 4 listed in Table 6.1 above. Hence the calculus fails to characterize its individuals and to assign them definite properties: it deals with unspecified individuals and attributes. Ergo it contains no specific laws, i.e., laws satisfied by objects of a definite kind, such as earthquakes or revolutions. In short the predicate calculus, by being semantically noncommittal, is attached to no ontology. But it is not an empty symbolism either: its small case letters are interpreted as unspecified individuals, its capital letters as unspecified predicates, and so on.

The indicated interpretation of logical calculi in terms of constructs of a certain type is the usual or *standard* interpretation but not the only possible one. Logical calculi can be assigned alternative interpretations – but then they may cease to be logical theories, i.e., theories concerned with deductive inference. A well known *nonstandard* model of the propositional calculus is the one in terms of switches in an electric network. And Kolmogoroff's interpretation of the intuitionistic propositional calculus in terms of problems is a nonstandard model of that calculus. These examples are mentioned only as a reminder that logical calculi are abstract theories except for the designation rules (e.g., ⌜'p' designates a proposition⌝), which we do not always care to state explicitly.

2. Mathematical interpretation

2.1. *Abstract Theory*

The mathematical theories occurring in factual science, such as trigonometry and the infinitesimal calculus, come with a definite mathematical interpretation. In other words they are specific ("concrete") theories concerned with mathematical objects of a definite kind, such as plane triangles or real functions. Thus the formulas ⌜$\sin^2 x + \cos^2 x = 1$⌝ and ⌜$d \sin x/dx = \cos x$⌝ are uniquely interpreted, namely in the field of real numbers. The latter can be extended to the field of complex numbers, but this is yet another specific structure: it is just an example or model of a field.

By contrast to such fully interpreted theories, those in logic, abstract algebra and topology are calculi with no fixed sense over and above the one determined by their axioms. These calculi, or *abstract theories* as we prefer to call them, are sometimes called *languages* or even *uninterpreted languages*. But this name seems misleading. First because, unlike a language, a theory, no matter how abstract, contains definite assumptions (axioms). Second because these assumptions give the theory a definite if parsimonious sense: it may be called the *minimal sense* of any theory built upon the given abstract theory by interpreting or specifying some or all of its concepts. Such a further interpretation will turn the abstract theory into a "concrete" theory with a richer sense – and correspondingly a narrower extension. The consideration of an example will clarify these points.

Consider lattice theory L. This is an abstract or formal theory concerned with a roomy structure $\mathscr{L} = \langle S, \preccurlyeq, \wedge, \vee \rangle$ that fits a number of species of specific mathematical objects. Being thus unattached to any specific interpretation, lattice theory can espouse (and subsequently divorce) a number of alternative interpretations. These are interpretations of a mathematical theory within mathematics: they are *mathematical interpretations*. And these are superimposed on the designation rules that turn a symbolism into an abstract theory – in this case L. A few such additional (or mathematical) interpretations of L are listed in Table 6.2.

TABLE 6.2

Four mathematical interpretations of lattice theory

Primitives of L	Order interpretation	Class interpretation	Propositional interpretation	Arithmetic interpretation
Abstract set S	Abstract set S	A collection F of abstract sets	The set P of propositions	The set N of natural numbers
Partial order $\preccurlyeq$	Partial order $\preccurlyeq$	Set inclusion $\subseteq$	Entailment $\vdash$	Divisibility $\cdot/.$
Binary operation $\wedge$	Greatest lower bound	Set intersection $\cap$	Conjunction &	Greatest common divisor
Binary operation $\vee$	Least upper bound	Set union $\cup$	Disjunction $\vee$	Least common divisor

Table 6.2 illustrates the following important points.

(i) Mathematical interpretation is a construct-construct relation and, more particularly, an *intertheoretical* affair. This contrasts with the other three kinds of interpretation listed in Table 6.1.

(ii) Mathematical interpretation is a *one-many relation* between the set of abstract theories and the set of "concrete" (specific) theories:

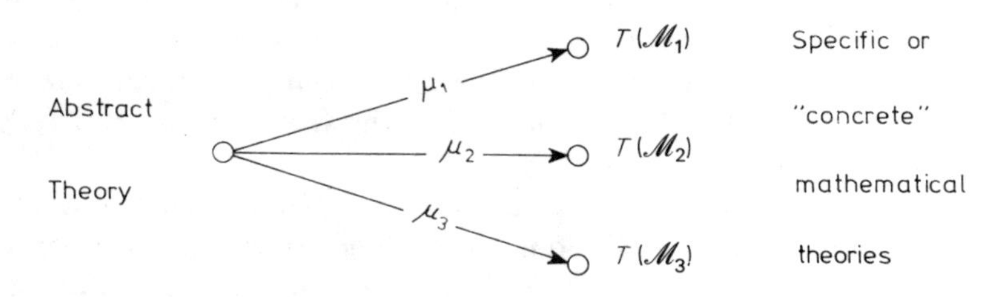

(iii) Not all of the interpretations of an abstract theory are equally "concrete" or *specific*. For example, the order interpretation of L maps S and $\preccurlyeq$ onto themselves and specifies only $\wedge$ and $\vee$. Thus the referents remain nearly as indeterminate as before. The class interpretation of L is more "concrete" or familiar but not completely so: the domain F could in turn be interpreted by specifying the nature of the sets in F. Only the propositional and the arithmetic interpretations are full, i.e., not susceptible to any further specification – except of course instantiation, as when a definite proposition is picked out of the set P.

(iv) Every specific structure, such as $\mathscr{F}=\langle F, \subseteq, \cap, \cup\rangle$ or $\mathscr{P}=\langle P, \vdash, \&, \vee\rangle$, is a realization or *model* of the abstract structure $\mathscr{L}=\langle S, \preccurlyeq, \wedge, \vee\rangle$. That is, the new (specific) constructs satisfy the formulas of the abstract theory L. Equivalently: the formulas of the abstract theory are satisfied in any of its models.

(v) A specific structure, or model, may be regarded as the value of an interpretation function μ that maps abstract primitives into specific ones. Examples:

$$\mathscr{L} \overset{\mu_1}{\mapsto} \mathscr{F}, \qquad \mathscr{L} \overset{\mu_2}{\mapsto} \mathscr{P}.$$

Let us take a closer look at models – in the mathematical or model theoretic sense of the word not in any of the alternative senses. (For the various senses of 'model' see Bunge (1973a).)

2.2. *Model*

An axiomatized theory may be construed as describing the structure constituted by its specific basic or primitive concepts. Thus the general theory T of partial order may be regarded as the theory about the abstract relational structure $\mathscr{A}=\langle S, \preccurlyeq\rangle$, where S is an arbitrary set and $\preccurlyeq$ an ordering of S. Since neither S nor $\preccurlyeq$ is definable in T, they are primitives of T. And since they are mutually independent, as well as sufficient to develop T provided some logic has been presupposed, $\mathscr{A}$ is the primitive base of T. Equivalently: T is the theory of $\mathscr{A}$, or $T(\mathscr{A})$ for short.

Let us emphasize the abstract character of $\mathscr{A}$. The elements of S are utterly faceless, hence $\preccurlyeq$ itself is quite anonymous except for the axioms of T that determine the sense of $\preccurlyeq$ – i.e., the properties of reflexivity, antisymmetry, and transitivity. This, then, is the basic purport or gist of T – that S is a partially ordered set. (Recall Ch. 5, Sec. 3.3.) It would be nonsense to say that T is meaningless. The axioms of T provide the *minimal sense* of any theory obtained by assigning to S and $\preccurlyeq$ a specific interpretation i.e., by exemplifying both primitives of T.

Take now any of the specific mathematical theories resulting from assigning both S and $\preccurlyeq$ definite senses within mathematics. Consider, in particular, the propositional model $\mathscr{P}$ and the real number model $\mathscr{R}$ of the abstract structure $\mathscr{A}=\langle S, \preccurlyeq\rangle$:

(I_1) $\mu_1(S)=$ The set P of propositions, $\mu_1(\preccurlyeq)=$ The relation $\vdash$ of entailment,

(I_2) $\mu_2(S)=$ The set R of real numbers, $\mu_2(\preccurlyeq)=$ The smaller than or equal to relation $\leqslant$.

The outcome of each interpretation of the primitive of T is a specific relational structure or *model*:

$$\mathscr{M}_1=\mathscr{P}=\langle P, \vdash\rangle, \qquad \mathscr{M}_2=\mathscr{R}=\langle R, \leqslant\rangle.$$

These are models or realizations of the abstract structure $\mathscr{A}=\langle S, \preccurlyeq\rangle$. Since the axioms of the abstract theory $T(\mathscr{A})$ are satisfied under either interpretation, they are said to hold (or to be true) in the corresponding model. Upon adjoining either interpretation assumption (or semantic formula) to T, we get a "concrete" (specific) theory i.e. one concerned with a definite species of objects, such as propositions or real numbers. The object of such an interpreted theory being a model or specific structure, we may call the former the *theory of the model*, or $T(\mathscr{M})$ for short.

In our case we have

$T(\mathscr{M}_1) = T(\mathscr{P}) = T(\mathscr{A})$ conjoined with the semantic assumptions I_1,
$T(\mathscr{M}_2) = T(\mathscr{R}) = T(\mathscr{A})$ conjoined with the semantic assumptions I_2.

$\mathscr{M}_1$ and $\mathscr{M}_2$ above are but two members of an unlimited population of models of $\mathscr{A}$. And they are all *full models* in the sense that they are obtained by interpreting every constituent of the abstract primitive base $\mathscr{A}$. We might also have built a family of partial models resulting from a *partial interpretation* of $\mathscr{A}$. It would be the family of all the structures in which the nature of S but not the one of $\leqslant$ is specified. (On the other hand it would be impossible to specify the order relation without at the same time fixing the nature of the elements of S.) In short, there are degrees of abstraction or, conversely, of semantic commitment. This notion is made more precise by

DEFINITION 6.1 Let $T(\mathscr{A})$ be an abstract theory with a primitive base $\mathscr{A} = \langle A_1, A_2, \ldots, A_n \rangle$ consisting of n nonlogical constants. Furthermore let $\mathscr{M} = \langle \mu(A_1), \mu(A_2), \ldots, \mu(A_n) \rangle$ be the value of an interpretation μ at $\mathscr{A}$. Finally, assume that μ does not effect a mere permutation (reshuffling) of the coordinates of $\mathscr{A}$. Then $\mathscr{M}$ is of *syntactic rank n*, *semantic rank* $m \leqslant n$ and *degree of abstraction* $\alpha = (n-m)/n =_{df} m$ of the interpreted primitives $\mu(A_i)$ differ from the corresponding abstract primitives A_i.

DEFINITION 6.2 $\mathscr{M} = \langle \mu(A_1), \mu(A_2), \ldots, \mu(A_n) \rangle$ is a *model* (or a *full model*) of $\mathscr{A} =_{df}$ the degree of abstraction of $\mathscr{M}$ is $\alpha = 0$. On the other hand, if $0 < \alpha < 1$, $\mathscr{M}$ is a *partial model* of $\mathscr{A}$.

Instead of the degree of abstraction α we might have defined the *degree of interpretation* $\beta = 1 - \alpha = m/n$. This would have had the advantage that it would not have involved the ambiguous term 'abstraction', which we use in its semantic acceptation not in the epistemological one of remoteness from sense experience. The concept of degree of interpretation will reappear in the theory of factual interpretation (Sec. 3.4).

We close this subsection with a couple of historical notes. The idea of a partially interpreted calculus, usually credited to Carnap (1939), goes back to Boole and was used by Whitehead (1898, pp. 10–11) in his campaign for the independence of algebra with respect to arithmetic. And the notion of a partial model introduced by Definition 2 must not be taken for the concept of semimodel introduced by Kemeny (1956): a semimodel involves full interpretation and differs from a model in that it does not involve validity in some structure.

2.3. *Intensional Models and Extensional Models*

We distinguish two kinds of mathematical interpretation and consequently two sorts of model: extensional and intensional. Equivalently: a model can be characterized either extensionally or intensionally. (Recall that our use of 'intensional' is the traditional one not the one current in modal logic.) An *extensional interpretation* assigns to every predicate in the abstract theory its extension in some field. For example, a binary relation is interpreted as the set of ordered pairs that stand in the given relation. On the other hand an *intensional interpretation* maps the abstract primitives into more specific mathematical objects that need not be set theoretic objects. For example, in the class interpretation of lattice theory considered in Table 6.2, Sec. 2.1, the lattice operations (meet and join) are assigned the class intersection and the class union respectively, and these operations are in turn characterized by the axioms of the algebra of classes.

More precisely, let T be an abstract theory formalized to the extent that all of its specific primitives can be identified and arranged in a sequence

$$\mathscr{A} = \langle A_1, A_2, \ldots, A_n, \ldots \rangle.$$

A possible *intensional model* of $\mathscr{A}$ is a structure

$$\mathscr{M}(\mathscr{A}) = \langle \mu(A_1), \mu(A_2), \ldots, \mu(A_n), \ldots \rangle,$$

the coordinates of which are definite mathematical objects with the same logical properties as their arguments in $\mathscr{A}$: an individual constant in $\mathscr{A}$ (e.g., the unit element in an algebra) is assigned an individual in $\mathscr{M}(\mathscr{A})$; a class in $\mathscr{A}$ is matched with a class in $\mathscr{M}(\mathscr{A})$; an m-ary relation in $\mathscr{A}$ is assigned an m-ary relation in $\mathscr{M}(\mathscr{A})$; a function in $\mathscr{A}$ is paired to a function in $\mathscr{M}(\mathscr{A})$, and so on. For example, in Table 6.2 we had the propositional interpretation of $\mathscr{L} = \langle S, \leqslant, \wedge, \vee \rangle$, to which it assigned the model $\mathscr{P} = \langle P, \vdash, \&, \vee \rangle$. Only the first coordinate of this quadruple is a set.

On the other hand a possible *extensional model* of an abstract structure $\mathscr{A}$ is produced by (*a*) introducing a nonempty domain of individuals D (the domain of the model) and (*b*) interpreting every coordinate of $\mathscr{A}$ as either a member of D or a set of m-tuples of members of D. In particular, an individual constant of $\mathscr{A}$ is matched with a member of D, an abstract unary predicate in $\mathscr{A}$ is interpreted as a subset of D, and every abstract m-ary predicate of $\mathscr{A}$ is assigned a subset of D^m. Every coordinate of a model is now a mathematical object with a definite set theoretic status: see Figure 6.1. In particular, the image of an abstract predicate is the

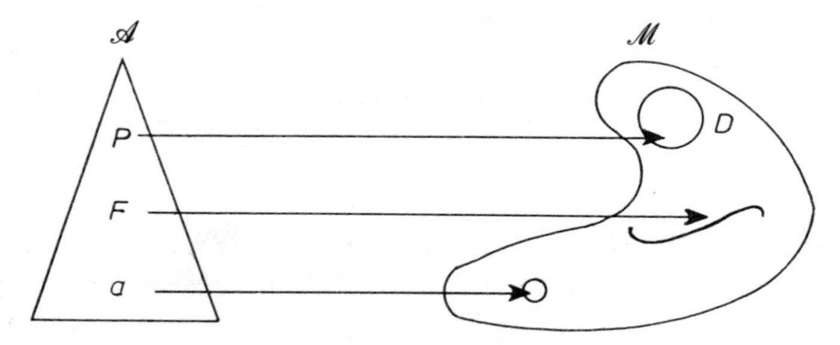

Fig. 6.1. An extensional interpretation maps the constituents of an abstract structure $\mathscr{A}$ into set theoretic objects built exclusively with the domain D of the model.

extension of the latter – not, as sometimes claimed, the meaning of the predicate.

The peculiarities of the two kinds of model are exhibited in Table 6.3 overleaf.

2.4. *Insufficiency of Extensional Models*

The discipline that studies extensional models (in the sense of Sec. 2.3) is called *model theory*. This is a sizable and growing chapter of metamathematics and it may be regarded as covering most of the semantics of logic and mathematics. The concern of model theory is "the mutual relations between sentences for formalized theories and mathematical systems in which these sentences hold" (Tarski, 1954, p. 572). For example, model theory investigates the relations between abstract Boolean algebra and its models. In particular, model theory can characterize the whole lot of models of a given abstract structure and can study the morphisms between such models.

Model theory is not only concerned with models *per se* but is also interested in the *use* of models to solve certain syntactic problems about

TABLE 6.3
Intensional model and extensional model

Abstract primitive	Intensional object	Extensional object
Individual constant a	Individual $\mu(a)$	$\mu(a)\in D$
Unary predicate P	Attribute $\mu(P)$	$\mathscr{E}(P)\subseteq D$
m-ary predicate P^m	m-ary attribute $\mu(P^m)$	$\mathscr{E}(P^m)\subseteq D^m$
m-ary operation or function F^m	m-ary operation or function $\mu(F^m)$	$\mu(F^m): D^m\to D$

any mathematical theory, whether abstract or specific. Indeed, model theory is the most powerful tool available to investigate questions of consistency, concept independence, definability, independence of axioms, provability, categoricity, etc. On this count model theory is relevant not only to pure mathematics but also to the foundations of science and to exact philosophy.

However, as conceived heretofore model theory is limited to extensional models and is therefore of restricted use even for purely mathematical purposes. Firstly, extensional models are not easy to come by: except in trivial cases sets are not given extensionally, i.e., by displaying their memberships, but are determined by some predicate or other. That is, normally a set is given by some law or rule which is not in turn resolvable in set theoretic terms. (Thus the fact that the *general* notion of a function can be partially elucidated as a set of ordered n-tuples does not entail that every *special* function can be so given. For example the logarithmic function is not given by a table of logarithms – the extensionalist ideal – but by certain formulas, such as $\ulcorner log(xy)=log\,x+log\,y\urcorner$, with $x, y\in R^+$.) In mathematics, just as in science, extensions are ultimately determined by senses. Secondly, even if it were possible to construct every model or example in purely set theoretic terms, intensional models could be dispensed with only provided we adopted the principle that coextensives are identical. But, as we say in Ch. 4, Sec. 1.2, this is a false dogma. It is particularly misleading with reference to factual science, where "descriptive interpretations" are essential (Carnap, 1958, p. 173). Therefore the claim (Suppes, 1961, 1967, 1969; Przelecki, 1969) that model theory can take care of the semantics of science is as unjustified as the identification of a model of an abstract structure ("formalized language") with "the real world" (Beth, 1962) or even with "a fragment of reality" (Przelecki, 1969).

Model theory does not tackle any of the problems peculiar to the semantics of factual science for the following reasons:

(i) The overwhelming majority of the mathematical theories used in factual science are not abstract but interpreted (within mathematics). Thus there is no way of reinterpreting a differential equation within mathematics: its degree of abstraction is nil. Now, model theory has little if anything to say about such theories – e.g., the theory of complex variables, the theory of integral equations, or differential geometry. Only abstract theories, such as group theory, or the general theory of topological spaces, pose model theoretic problems such as "Does this interpretation of the primitives yield a model?", "Are all the models of the given structure isomorphic to one another?", or "Can we prove a representation theorem for this theory?".

(ii) The models occurring in "intuitive" (nonformalized) mathematics and in science are mostly intensional models, i.e., they are "defined" by properties and laws rather than extensionally. On the other hand the models studied by model theory are extensional, hence incapable of discerning intensional differences unless the latter are accompanied by extensional differences. Applied mathematics and science cannot dismiss such differences, particularly since coextensive predicates may be characterized by different law statements, whence they must be counted as distinct.

(iii) As practised in formalized mathematics, which is the object of model theory, axiomatics involves *de*interpretation. For example, the abstract theory of natural numbers is formulated in such a way that the very concept of a natural number is not explicitly included in it, precisely in order to allow for alternative interpretations. A possible axiomatization of this abstract theory boils down to the following set of postulates:

A1 $x' \neq 0$.
A2 $x' = y' \Rightarrow x = y$.
A3 $(P0 \,\&\, (Px \Rightarrow Px')) \Rightarrow (y)\, Py$.

One recognizes here the kernel of the five Dedekind-Peano axioms. But the above formulas are satisfied in models other than the standard number theoretic one. In order to have the above postulates describe the essential properties of the natural numbers, suitable interpretation assumptions must be joined to them. Interpretation is thus external to formal axiomatics as opposed to the axiomatics found in "concrete" or "intuitive" mathematics and in science. (For an elucidation of the differences between formal axiomatics and *inhaltliche Axiomatik* see Hilbert-Bernays, 1968, I, Sec. 1.) In particular, scientific axiom systems must contain interpretation assumptions, as emphasized by Carnap (Carnap, 1939, 1958). Otherwise we would not know what the theory is about nor, consequently, how to apply and test it.

(iv) Because science concerns the external world, scientific theories must involve not only mathematical interpretations but also factual ones, i.e. construct-fact correspondences. These correspondences fall outside the scope of model theory, which is concerned with mapping structures into structures. The semantic assumptions in factual science correlate definite mathematical structures with real systems – and a real system is not a mathematical object. (The fashionable identification of models with possible worlds has suggested the view that the actual world is just a possible model. This new fangled version of the Platonic allegory of the cavern overlooks a couple of details. One is that, while a model is a harmless unpolluted construct, the world is not the work of a mathematician. Another is that, while a formula may or may not be satisfied in a model, the natural laws are inherent in the real world. A third is that, while every single model is fully characterized, no chunk of reality, however minute, is known exhaustively.) Furthermore the semantic assumptions in factual science are refutable hypotheses (Ch. 3). For example, improved accuracy in measurement showed that Yukawa's theory was not about μ-mesons, as originally conjectured, but about π-mesons. In contrast, the rules of assignment (of extensions) given by an extensional model may be regarded as holding analytically provided analyticity is construed in a permissive fashion (Kemeny, 1956).

In sum, model theory does not help us to elucidate the semantic peculiarities of factual science. The semantics of science takes off where model theory leaves: see Figure 6.2.

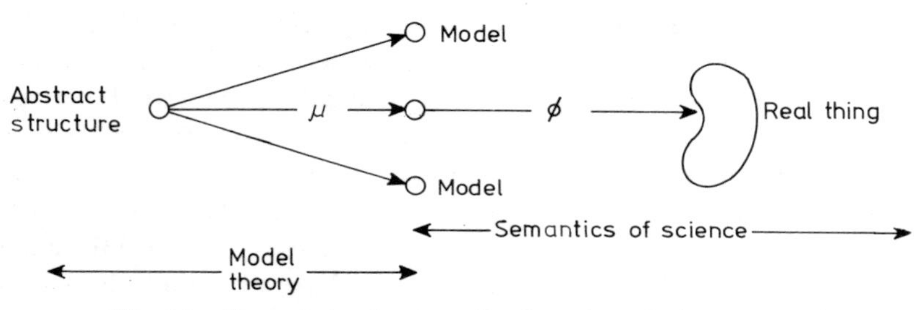

Fig. 6.2. From abstraction to reality through model (or conversely).

Let us now turn to the problem of factual interpretation – the map ϕ that model theory does not deal with.

3. Factual interpretation

3.1. *The Need for Factual Interpretation in Science*

Every contemporary scientific theory worth the name has a mathematical formalism. This formalism consists of a set of mathematical theories the symbolism of which is interpreted by tacit or explicit designation rules pairing symbols to constructs. The overwhelming majority of such formalisms are not abstract theories but theories of models in the sense of Sec. 2.2. Probability theory and the theory of Hilbert spaces are theories of this kind and the two are components of the mathematical formalism of quantum mechanics. Whatever additional interpretation any such formalism be assigned, is nonmathematical – hence it cannot be described exclusively in mathematical terms. Thus a physical geometry will consist of a mathematical geometry together with semantic assumptions pairing constructs to things or properties of things.

However, many mathematicians see no need for such additional interpretations. Thus a distinguished mathematical physicist postulates that the isotropic subspaces of R^n are light rays, whence ray optics is just the theory of such spaces (Jost, 1965, p. 18). Another distinguished mathematician proposes the following definition: "A lever is a system consisting of a plane π, a straight line ℓ in that plane, called a beam, a point O on that line called the fulcrum, etc." (Freudenthal, 1971, p. 316). Mind: in the two quotations certain mathematical objects are not said to *represent* certain physical objects but are *identified* with them. Finally an eminent philosopher-scientist has forcefully defended the slogan "To axiomatize a theory is to define a set-theoretical predicate", i.e., "one that can be defined within set theory in a completely formal way" (Suppes, 1967). If this is how scientific theories are to be reconstructed then it is obvious that (*a*) logic applies to physical objects such as dynamical systems: "as far as dynamical systems are concepts (a lever, a solar system) they allow for logical relations" (Freudenthal, 1971, p. 321); and (*b*) "there is no theoretical way of drawing a sharp distinction between a piece of pure mathematics and a piece of theoretical science" (Suppes, *op. cit.*). The view, held by all three authors, that a scientific theory consists of its mathematical formalism alone, may be regarded as an updated version of the Pythagorean philosophy and may be called *semantic formalism* – or *unsemantics* for short.

Most theoretical scientists are not semantic formalists: they hold with Einstein (1936) that a scientific theory has a content that overflows its mathematical formalism and thereby throws the theory at the mercy of facts. Suppes himself acts on this nonformalist conviction when expounding scientific theories. Thus he formulates his theoretical model of individual decision in the following fashion (Suppes, 1969, p. 148). "We shall call an ordered triple $\mathscr{S} = \langle S, C, D \rangle$ an *individual decision* situation when S and C are sets and D is a set of functions mapping S into C. The intended interpretation is:

S = set of states of nature,

C = set of consequences,

D = set of decisions or actions."

These interpretation assumptions are not included in the axiomatic definition of an individual decision situation but are juxtaposed to the latter: nevertheless they are not forgotten although they certainly are not set theoretic constructs.

The typical theoretical scientist will not say that a factual item f (thing, property, state, event, process) *is* a mathematical object m but rather that *m represents f*. He knows that one and the same mathematical object (set, function, space, equation, etc.) may be assumed to represent different factual items in different theories. For example, the Laplace equation occurs in at least the following capacities:

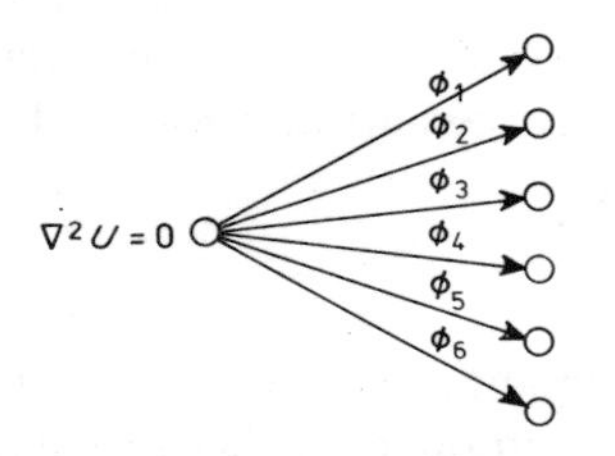

Velocity field of incompressible fluid
Static gravitational field in a vacuum
Electrostatic field in a vacuum
Magnetostatic field in a vacuum
Stationary temperature distribution
Atomic state for zero energy level

Since scientists meet the same functions and equations over and over again in different chapters and associated with different factual contents (senses and referents), they know that a scientific theory has a factual content which is not exhausted by its formalism. Most scientists realize that whatever can be read *out* of a mathematical formalism is what we have more or less unwittingly read *into* it. They differ only as to the nature of this content and the way it should be assigned. Thus while most scientists seem to favor a realist but sloppy semantics, those who do take pains to spell out the interpretation assumptions do it often in operationist terms. Only a few hold the magic view that a mathematical formalism yields its own interpretation (Everett, 1957; DeWitt, 1970). To help settle these problems let us analyze a couple of examples.

3.2. *How Interpretations are Assigned and What They Accomplish*

Let us exhibit three examples of factual interpretation with a view to finding out what it adds to the mathematical formalism.

Theory 1: Rat race

Primitives: S, Δ.

Axiom 1 Δ is a binary associative operation in the set S: i.e., for any $x, y, z \in S$, $(x\Delta y)\,\Delta z = x\Delta\,(y\Delta z)$.

Axiom 2 $\phi(S)$ = collection of rats.

Axiom 3 $\phi(\Delta)$ = joining a race, i.e., $\phi(x\Delta y) = x$ joins y (in this order) in a race to some goal.

The mathematical formalism boils down to Axiom 1, which states that

$\langle S, \Delta \rangle$ is a semigroup. The other two postulates are semantic assumptions. By virtue of the latter, the former becomes the law statement ϕ (A1)=⌜First to enter race wins⌝ (or ⌜Late comers are eliminated⌝).

Without A2 and A3 no such "reading" (interpretation) of A1 would be explicit and unambiguous. The first semantic assumption identifies the referents of the theory as rats. The second stipulates that Δ represents race joining. Axiom 1 determines the mathematical sense of Δ, but the full sense of Δ is given by all three axioms.

Theory 2: Switching circuits

This theory contains half a dozen semantic assumptions defining an interpretation function ϕ that relates certain formulas to elements of circuitry of a kind. This function is a one to one mapping $\phi: B \to N$ from the set B of Boolean functions of a certain type onto the set N of two terminal series-parallel electrical networks. The domain of ϕ is a construct while its range is an aggregate of pieces of hardware: ϕ is a factual interpretation function. Given any Boolean form b in B, ϕ locates a possible network n in N such that $\phi(b)=n$, i.e. such that *b represents n.* And conversely: given a possible network n, its Boolean image will be $b=\phi^{-1}(n)$, where ϕ^{-1} is the inverse of ϕ. *Example*: Figure 6.3.

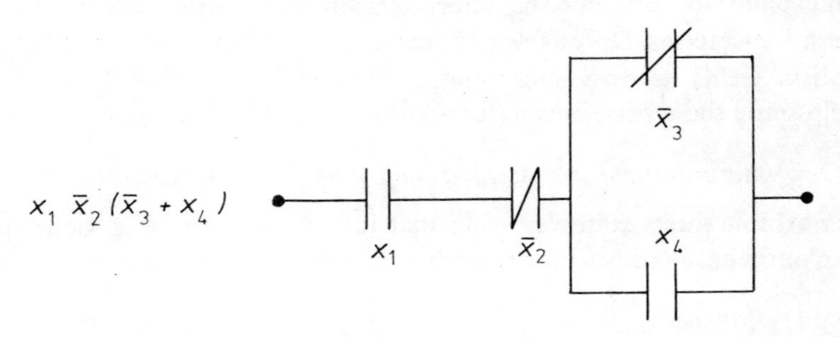

Fig. 6.3. A representation of switching circuits by Boolean forms.

The specific semantic assumptions determining ϕ are (Harrison, 1965, p. 79)

A1 $\phi(0) = \circ$ (Equivalently: $0 \hat{=} \circ$)

A2 $\phi(1) = \circ\!\!-\!\!-\!\!-\!\!\circ$ (Equivalently: $1 \hat{=} \circ\!\!-\!\!-\!\!-\!\!\circ$)

A3 For any variable x_i, $\phi(x_i) = \circ\!\!-\!\!\!-\!\!\!\dashv\ \vdash\!\!\!-\!\!\!-\!\!\circ$ x_i

(Equivalently: $x_i \hat{=}$ a normally open contact)

A4 For any variable x_i, $\phi(\bar{x}_i = \circ\!\!-\!\!\!-\!\!\!\not|\!\!\!-\!\!\!-\!\!\circ$ x_i

(Equivalently: $\bar{x}_i \hat{=}$ a normally closed contact)

A5 For any Boolean forms a, b in B:

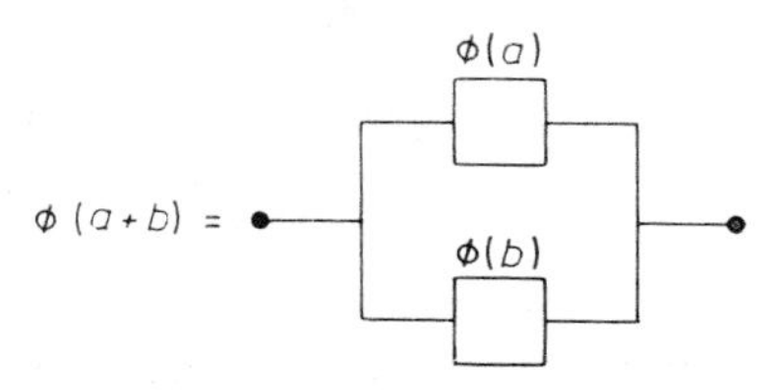

(Equivalently: $a + b \hat{=}$ two terminal parallel circuit.)

A6 For any Boolean forms a, b in B:

$\phi(ab) =$ •—□—□—•

(Equivalently: $ab \hat{=}$ two terminal series circuit.)

Here again the semantic assumptions determine both the reference class and the way the constructs represent some of the features of their referents.

Theory 3: Assembly theory

The preceding considerations apply not only to scientific theories but also to theories in scientific or mathematical metaphysics, such as assembly theory (Bunge, 1971b). This theory is concerned with the basic modes of assembly or composition of systems apart from their specific properties. It may be regarded as ring theory (a stalwart member of abstract algebra) together with the following semantic assumptions

A1 $\phi(S) =$ the set of all systems
A2 $\phi(0) =$ the null system
A3 $\phi(+) =$ system juxtaposition or joining
A4 $\phi(\cdot) =$ system interpenetration or superposition

By virtue of these semantic assumptions every formula in ring theory becomes a metaphysical statement. For example, for any x, y, z in S,

$\phi(x(y+z)=xy+xz)$ = The superposition of system x with the outcome of the juxtaposition of systems y and z equals the juxtaposition of the systems (x superposed to y) and (x superposed to z).

We now generalize the preceding considerations by defining a theoretical factual construct as a mathematical construct together with a factual interpretation map. More precisely, we adopt

DEFINITION 6.3 A construct c will be said to be a *theoretical factual construct* iff

(i) c belongs to a theory and

(ii) $c=\langle m, \phi\rangle$, where m is a mathematical construct and ϕ is an interpretation map such that $\phi(m)$ is a factual item (thing, property, or event) or a collection of factual items.

Example The ordered pair $\langle M, \phi\rangle$ is the concept of mass in particle mechanics iff $M: P \to R^+$ is an additive function and

(i) $\phi(P)$ = Particles,

(ii) $\phi(M(x))$ = Inertia of x for every $x \in P$,

(iii) M occurs in the equations of motion of particle mechanics multiplying the particle acceleration.

To conclude we collect the lessons learned from our analysis:

(i) The nonsemantic axiom(s) of a theory determine(s) the mathematical sense of its primitives;

(ii) the semantic axiom(s) determine(s) the referents and sketch(es) the full factual sense of the primitives and of the nonsemantical axiom(s);

(iii) the sense and reference of the derived constructs of a theory are determined by its axioms.

In sum, the sense and reference of a theory are determined jointly by all of its axioms. Equivalently: the significance of the symbolism ("language") of a theory is given by all of its axioms taken together.

3.3. *The Factual Interpretation Maps*

The first three examples discussed in the last subsection are extremely simple, hence atypical: they involved a single interpretation. In fact, in each of them the interpretation function ϕ mapped an abstract structure $\mathscr{A}$ into a factual domain $\mathscr{F}$. Thus in the first case $\mathscr{A}=\langle S, \Delta\rangle$, and $\mathscr{F}$

consisted of a collection of racing rats. There was no intermediary mathematical model such as, say, the ring of integers or Euclidean geometry. In brief, we had

$$\mathscr{A} \overset{\phi}{\mapsto} \mathscr{F}. \tag{1}$$

Moreover in the second case ϕ was no less than an isomorphism between the set B of constructs and the collection N of things. Besides, in this case as well as in the case of assembly theory ϕ was a morphism of addition and of multiplication. Such simplicity is exceptional in science.

In most scientific theories the domain of ϕ is not an abstract structure but a model of such. In other words, the mathematical formalism of the typical scientific theory is a *theory of a model.* And this theory is seldom found ready-made in a mathematical rack: the theory is usually built by enriching an interpreted mathematical theory (or rather a motley collection of fragments of interpreted mathematical theories) with some specific assumptions not found in mathematics. For example, a classical field theory is obtained by putting the following components together: (*a*) the theory of differentiable manifolds, (*b*) a set of specific formulas, chiefly the field equations, boundary conditions, and constraints, and (*c*) a set of semantic assumptions.

In these cases we have two successive and dovetailing interpretations: μ and ϕ, the former from an abstract structure $\mathscr{A}$ to a model $\mathscr{M}$, the latter from $\mathscr{M}$ to a factual domain $\mathscr{F}$:

$$\mathscr{A} \overset{\mu}{\mapsto} \mathscr{M}, \qquad \mathscr{M} \overset{\phi}{\mapsto} \mathscr{F}. \tag{2}$$

Hence, with a qualification to be mentioned shortly, we may regard the factual interpretation of an abstract structure as the composition of the two mappings, i.e. $\phi \circ \mu : \mathscr{A} \to \mathscr{F}$. (Bunge, 1972b.) It is true that the abstract structure is seldom if ever dug out when analyzing a typical scientific theory: one usually starts from a model. However, we do not get the full semantical picture unless we uncover that deeper layer.

In point of fact only a portion $\mathscr{M}_0$ of a mathematical model is usually assigned a factual interpretation. For example, not every vector decomposition, or series expansion, or integral representation, nor even every solution of a differential equation is always assigned a factual counterpart. A part of the mathematical formalism of a factual theory is usually either idle or plays a purely syntactical role. (For example, in Theory 3 of Sec. 3.2 the unit of the ring is assigned no special interpretation.) Consequently ϕ is normally a *partial* function from $\mathscr{M}$ to $\mathscr{F}$. Equivalently: ϕ is a total function on a *sub*set $\mathscr{M}_0$ of $\mathscr{M}$. We indicate this by writing

$$[\mathscr{M}] \overset{\phi}{\mapsto} \mathscr{F} \tag{3}$$

and drawing Figure 6.4.

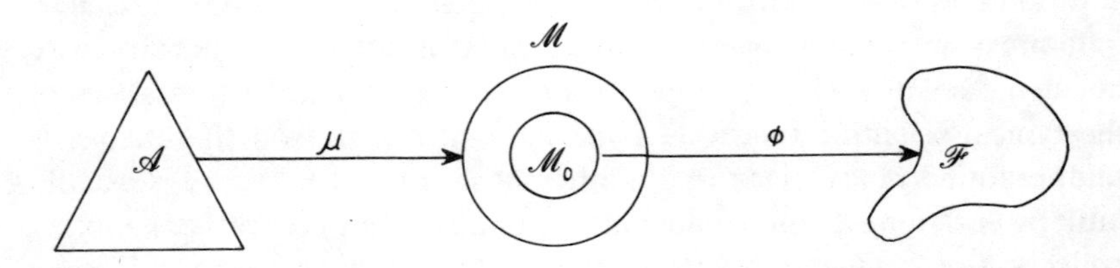

Fig. 6.4. Normally a factual interpretation ϕ maps only a part $\mathscr{M}_0$ of a model $\mathscr{M}$ of an abstract structure $\mathscr{A}$ into a factual domain $\mathscr{F}$.

In the case of the switching circuits theory discussed in Sec. 3.2 the interpretation map ϕ was one to one and consequently ϕ^{-1} had an inverse. That is, two constructs were the same (different) just in case their factual images were identical (different). We warned already that this is the exception rather than the rule: usually ϕ does not discern among equivalent systems. Actually this is so even in the case of the theory of switching circuits, which does not distinguish among circuits constructed with different materials and having different sizes as long as they are topologically equivalent. That is, ϕ is really a mapping of $\mathscr{M}_0 \subset \mathscr{M}$ into a *family of equivalence classes of concrete systems*. In other words, the range of ϕ is not a factual domain $\mathscr{F}$ but the quotient of $\mathscr{F}$ by an equivalence relation $\sim$, i.e. $\mathscr{F}/\sim$. This equivalence relation is defined in a tacit fashion by the very theory T in question, namely thus: Two factual items are *equivalent with respect to* T iff T does not discern between them, i.e. iff it represents them by the same constructs. In brief, instead of (3) we usually have

$$\phi : [\mathscr{M}] \to \mathscr{F}/\sim, \tag{4}$$

where $\mathscr{M}$ is a collection of mathematical models and $\mathscr{F}$ a set of factual domains.

In sum, we distinguish four different factual interpretation maps ϕ:

Abstract theory → Factual systems (1)
Theory of a model → Factual systems (2)
Part of a theory of a model → Factual systems (3)
Part of a theory of a model → Equivalence classes of factual systems. (4)

In either case we adopt the following

DEFINITION 6.4 If $\mathscr{M}$ is a mathematical model and ϕ a factual interpretation map, then the structure $\mathscr{M}_\phi \doteq \langle \mathscr{M}, \phi \rangle$ is called a *factual model.*

DEFINITION 6.5 If $\mathscr{M}_\phi$ is a factual model, then a theory $T(\mathscr{M}_\phi)$ of such a model is called a *factual theory*.

The preceding elucidations suffice to settle certain much debated questions such as whether gravitation theory is reducible to geometry, as is often held with reference to Einstein's general relativity. If the semantic assumptions of the theory are not stated explicitly, then of course one has, strictly speaking, only a mathematical formalism not a factual theory. However, the context makes it usually clear what things (material systems and fields) the theory is about and which of the properties (e.g. gravitational interaction) it represents. That is, the formalism is treated as a full fledged factual theory. Moreover, and this is a strictly syntactic point, if gravitation theory were just a geometrical theory, then it would contain no formulas other than those of Riemannian geometry – but in fact it adds to the latter its own field equations and equations of motion.

3.4. *Factual Interpretation: Full and Partial*

We saw in the preceding subsection that the mathematical formalism of a theory is likely to contain components without representatives in the real world. For example, not every Fourier analysis will represent the spectral decomposition of a wave packet. Conversely, not every trait of a factual system is likely to be represented by any theory of it. Thus half of the wave-like solutions to Maxwell's field equations are usually discarded because they would have to be interpreted as representing waves coming from the future. (They are often called 'unphysical' or 'physically meaningless' solutions.) On the other hand the same equations fail to account for the photon structure of a light beam. Briefly: the theory has redundant constructs and also leaves some factual items dangling in the midst of reality. This is quite general: we may assume that no $\mathscr{M}$ is isomorphic with $\phi(\mathscr{M})=\mathscr{F}$ except perhaps in some limited respects. Usually $\mathscr{M}$ contains elements with no images in $\mathscr{F}$ and, conversely, $\mathscr{F}$ has elements not accounted for by $\mathscr{M}$. See Fig. 6.5.

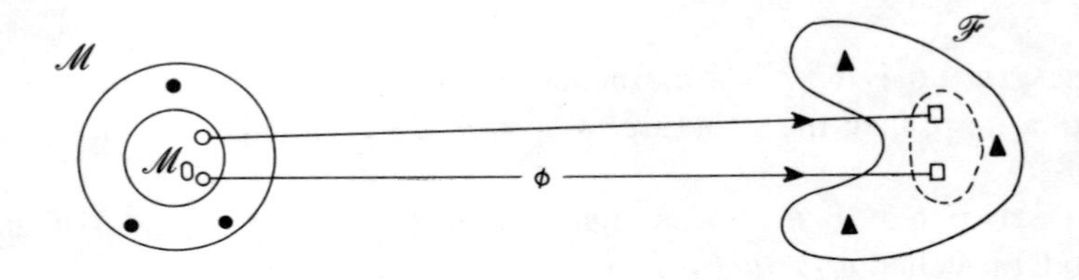

Fig. 6.5. Black dots = redundant constructs. Black triangles = orphan entities.

From a semantic point of view the best scientific theory referring to a given factual field is the one with the fewest black spots and leaving the fewest black triangles. This holds for the formulas (e.g., equations) and their constituents (e.g., parameters in the equations). Rich theories are likely to contain idle formulas but on the other hand are bound to contain few if any uninterpreted parameters, while shallow theories abound in such parameters. (A *fundamental theory* is often defined as containing no constants other than universal constants.) While a redundant formula can be isolated and immobilized, the factually uninterpreted parameters cannot be disposed of short of replacing the theory. Such parameters are of the essence of phenomenological (or black box) theories as well as of the hypotheses that cover the available data and little else. Since those parameters can be varied almost *ad libitum* to conform to the data, the corresponding theory is a docile data recipient with only a weak explanatory power (Bunge, 1963b, 1964, 1967a). The more detailed a "picture" of reality a theory supplies, the more heavily interpreted it is: the less specific, the less semantically committed.

The degree of semantical commitment of a scientific theory can be quantified with the help of the Definition 1, in Sec. 2.2, of abstraction degree. We presently adapt it to a factually interpreted model and a factually interpreted theory as characterized by Definitions 4 and 5 in Sec. 3.3:

DEFINITION 6.6 Let $\mathcal{M}=\langle M_1, M_2, \ldots, M_n\rangle$ be a mathematical model of some abstract structure and let ϕ be a factual interpretation of $\mathcal{M}$. If $m \leqslant n$ of the interpreted concepts $\phi(M_i)$ for $1 \leqslant i \leqslant n$ differ from the corresponding mathematical primitives M_i, then the factual model $\mathcal{M}_\phi = \langle \mathcal{M}, \phi\rangle$ and any theory $T(\mathcal{M}_\phi)$ of it are said to possess *a degree of interpretation* $\beta = m/n$.

DEFINITION 6.7. Let $\mathcal{M}_\phi = \langle \mathcal{M}, \phi\rangle$ be a factual model and $T(\mathcal{M}_\phi)$ be a theory of $\mathcal{M}_\phi$. Then

(i) $\mathcal{M}_\phi$ and $T(\mathcal{M}_\phi)$ are said to be *fully interpreted* in factual terms iff $\beta = 1$.

(ii) $\mathcal{M}_\phi$ and $T(\mathcal{M}_\phi)$ are said to be *partially interpreted* in factual terms iff $0 < \beta < 1$.

Example 1 Probability theory is based on two primitives: the sample space S and the probability measure P. If neither is assigned a factual interpretation, $\beta = 0$ and the theory remains in pure mathematics. If the elements of S are interpreted as states of a system or as events of some kind, e.g. learning a certain task, then $\beta = \frac{1}{2}$. Most stochastic learning theories are of this kind, i.e. semantically half committed. Finally, if also P is interpreted, then $\beta = 1$. For example, if $s \in S$ is assigned an event of a certain kind, then $P(s)$, which is some number in the real number interval $[0, 1]$, might be interpreted as the tendency or disposition for that event s to occur. This would complete the interpretation of the stochastic theory concerned – without of course ensuring its truth. (On the other hand identifying $P(s)$ with the relative frequency of s would not count as an interpretation of probability but as an estimate of probability values.)

Example 2 Let $\mathcal{M} = \langle S, F, G, k\rangle$ where S is a set, F and G real valued functions on S, and k a positive real. So far this is a specific structure or model. Now introduce an interpretation map ϕ such that

$$\phi(S) = \text{set of bodies},\ \phi(F) = \text{mass},\ \phi(G) = \text{volume},\ \phi(k) = k.$$

We get a factual model $\langle \mathscr{M}, \phi \rangle$ with degree of interpretation $\beta = \frac{3}{4}$. And if a different map ϕ' is chosen, one that assigns to every coordinate in $\mathscr{M}$ a factual item, then β jumps to 1.

3.5. *Generic Partially Interpreted Theories*

A *generic factual theory* is one concerned with a genus rather than a species of factual items – e.g., with bodies, or organisms, or societies. When adjoined specific assumptions it can become a specific theory prepared to deal with certain traits of some species of systems – say fluids, or flat worms, or industrial societies. In short, the reference class of a generic theory in factual science is some genus or other – not necessarily a natural genus. And what it represents are some pervasive features of its referents.

There are two kinds of factual generic theory: fully interpreted and partially interpreted in factual terms. The giant classical theories of science, such as classical mechanics and the theory of evolution, are fully interpreted generic theories: they are concerned with whole families of species and all their basic concepts are assigned a factual interpretation. In addition to these theories there are generic factual theories with a low degree of interpretation. This lack of a firm semantical commitment makes them readily portable from one field of research to another. The first outstanding specimen of this kind of theory was Lagrangian dynamics, which started out as a chapter of mechanics, then spread over almost all of physics (cf. Bunge, 1957, 1967b) and has now found its way into the general theory of systems of any kind (White and Tauber, 1969). Newer members of the species are information theory, the mathematical theory of machines, and general network theory. All of these theories are factual in the sense that they refer to real systems though not to any definite species of such. And, far from representing specific properties, they represent only very general features. Consequently they are useful guides in either of the following situations: (*a*) absence of detailed knowledge of the system; (*b*) a detailed knowledge may be available but only certain outstanding features compatible with a number of possible mechanisms are of interest, and (*c*) a unified treatment of a number of research subjects is intended – for instance in order to bring out their common formal traits.

To get a feel of the semantic peculiarities of such generic semi-interpreted theories in factual science, as well as of the methodological problems they raise, we shall take a look at the Rashevsky-Turing theory of morphogenesis (Rosen, 1970, I, Ch. VII). This theory states, in a nutshell, that any initially homogeneous or amorphous system that reaches an unstable state may evolve towards a final state of inhomogeneity (e.g., polarity) under the action of slight external perturbations. The state variables of this theory are left factually uninterpreted. Only the independent variable is interpreted, namely as time. Moreover, the theory is kinetic rather than dynamic, in the sense that it assumes no specific forces or interactions responsible for the process: any force will do as long as it is compatible with the equations of change of state. In sum, the Rashevsky-Turing theory of morphogenesis is a morphological theory: a theory of the genesis of differentiation or form in nearly any complex system. It is richer than a black box theory in that it accounts for certain changes in the interior of the box, but is equally noncommittal regarding the nature of the components and their interactions.

A Rashevsky-Turing system is defined as any system satisfying the assumptions of the theory, whatever its actual physics and chemistry may be. In other words, the Rashevsky-Turing theory has a number of possible interpretations. It is not just that it concerns a whole class of systems – every general theory does. It concerns a family of classes, i.e. a genus. As soon as the state variables of the theory are specified, i.e. as soon as they are assumed to represent definite properties, interactions, etc., one species in the genus is singled out for reference. In short, upon interpreting the state variables of the theory, the family of species is restricted to a single species of morphogenetic systems.

This semantic difference between a partially interpreted and a fully interpreted theory is of importance for methodology. Since a generic morphogenesis theory specifies neither the substratum nor the forces acting in it, no definite predictions can be computed with the formulas of the theory. Hence such a theory is untestable in the usual way. Partially interpreted theories call for a revision of the conventional methodology of science. In fact such theories are tested vicariously, namely by testing some of the specific theories obtained upon specifying (interpreting) the state variables as definite properties of a system of a definite kind (Bunge, 1973a, Ch. 2).

3.6. *Principles of Factual Interpretation*

Any given mathematical construct may be assigned a number of different factual interpretations. Therefore there is often uncertainty, and occasionally spirited controversy, as to which interpretation is the best. Consequently it is desirable to have a battery of explicitly formulated criteria for an admissible factual interpretation – if not to facilitate the interpretation task at least to ease rational argument about it. We propose the following conditions that a sound interpretation of a mathematical construct in factual terms should satisfy: the interpretation should concern a reasonably safe mathematical construct; it should not originate inconsistencies; it should be strict, i.e. adjusted to the mathematical formalism; it should be literal not figurative; it should be factual rather than empirical; it should be full not partial; and it should aim at the truth. Let us spell out these conditions.

(i) *Factual interpretations should bear on mathematically sound formalisms.* If the mathematical skeleton is ambiguous or inconsistent, no amount of cunning interpretation will turn it into a reasonable factual theory. This seems self-evident, yet some highly refined scientific theories, such a quantum electrodynamics, fail to satisfy the condition: they contain ambiguous expressions (e.g., integrals whose value depend on the mode of computation) and inconsistencies (e.g., the electric charge, assumed to be finite when occurring in an equation of motion, turns out to be infinite in derived formulas). Hence the interpretation of such theories must be regarded as insecure. And instead of trying to save the sick formalism by a semantic *tour de force* one should try alternative formalisms. But before anyone attempts to bell this cat the dogma that quantum electrodynamics is perfect will have to be shaken.

(ii) *Factual interpretations should introduce no inconsistencies.* There is danger of inconsistency whenever a construct is assigned different factual correlates – i.e. if the theory contains more than one interpretation map. However, this is sometimes necessary and need not lead to any inconsistencies. For example, a neuropsychological theory may contain variables that are assigned both a neurological and a psychological interpretation. Thus in Grossberg's theory of learning networks every vertex function is interpreted both as a stimulus trace and as an average membrane potential (Grossberg, 1969). These happen to be mutually compatible interpretations of one and the same mathematical construct. On the other hand the standard formulations of quantum mechanics contain multiple interpretations that do lead to contradiction – as when the 'Δx' occurring in Heisenberg's inequalities is interpreted as both the average scatter of the *particle* position and the width of the *wave* packet, and perhaps also as the physicist's *uncertainty* concerning the exact particle position. (See Bunge, 1973b.)

(iii) *Factual interpretations should be strict not adventitious.* A factual interpretation should match the structure of the construct concerned: it should pour no more content than the construct can possibly hold. For example, if a hamiltonian contains only variables referring to a given system (e.g., a molecule) then it should not be taxed with representing both the system and an unspecified measuring device – let alone the experimenter's mind. In general, the value of a function should not be interpreted as concerning more referents than arguments. If it be assumed that a formula α concerns a certain fact f, then α should contain at least one variable x such that $\phi(x)=f$. Otherwise it must be concluded that the interpretation is adventitious: that it has no mathematical leg to stand on (Bunge, 1969).

(iv) *Factual interpretation should be literal not metaphorical.* In mathematics the concept of analogy can be elucidated in an exact way, namely as homomorphism, and can thus be kept under control. Outside mathematics analogy wears too many faces, all of them blurred and tinged by subjectivity: one man's similarity is another's dissimilarity. Metaphor can be a pragmatic asset: it may have heuristic value and may also be of some use in teaching but it can be terribly misleading precisely for being highly subjective. For this reason it does not belong in scientific theory – *pace* a rather fashionable view (Black, 1962; Hesse, 1965). The aim of a new scientific theory is not to win neophobic followers but to account for things with characteristics of their own – which traits the metaphor is bound to hide, as the gist of metaphor is to pass the new for old. A scientific theory must involve literal interpretations only – no *as ifs*. This need was first realized at the turn of the century when

Maxwell's electromagnetic theory was freed from any mechanical associations, and is now acutely felt in relation with quantum mechanics. Indeed the classical analogies of position and momentum, and particle and wave, though probably inevitable in the early stages, have introduced inconsistencies and have blocked the understanding of the theory as an original creation referring to *sui generis* things (Bunge, 1967c). In short: analogy belongs in the scaffolding not in the building. (Metaphor is dead but it won't lie down.)

(v) *Scientific theories should be interpreted by reference to facts not test procedures.* For example, acidity is indicated by, but not interpretable as, a color change in litmus paper; and physiological stress is not interpretable as organ enlargement as revealed by autopsy. If on the other hand an analysis of observation (or of measurement) is held to be the key to the factual interpretation of a theory, then (*a*) meaning is being mistaken for testability and (*b*) the scope of the theory is restricted to situations under experimental control. Which is what happens with the standard or Copenhagen interpretation of quantum mechanics. The result is not only confusion but also inconsistency, as exemplified by Bohr's thesis that the theory, though nonclassical, is based on classical physics (i.e., presupposes the latter) just because the end results of measurements are describable in classical terms. If operationism is given up then the quantum theories can be interpreted in their own revolutionary terms – as demanded, albeit timidly, by Wheeler (1957) and Everett (1957) – as well as in strictly objective terms (Bunge, 1967b).

(vi) *Factual interpretations should be global not spotty.* Not isolated formulas but whole formalisms should be assigned a factual interpretation if the risks of inconsistency and irrelevance are to be avoided. Taken in isolation any formula can be interpreted in a number of ways: taken together with other formulas of the same conceptual system, the number of interpretations is decreased because the number of conditions to be satisfied is increased. For example, Shannon's formula for the quantity of information (cf. Ch. 4, Sec. 3.2) looks like Boltzmann's formula for the entropy and is therefore often interpreted as the entropy of the system. However, this metaphoric interpretation is quite arbitrary, as the information theoretic "entropy" is not related to any thermodynamic function such as energy, temperature, pressure, or volume. Consequently it is not entitled to be called 'entropy' – nor is entropy entitled to pass for quantity of information. Just as mathematical interpretation is constrained by the requirement that it yield formulas satisfiable in some model, so factual interpretation should produce formulas reasonably true to fact – in particular it should ensue in statements representing laws. In other words, factual interpretation is not a matter of convention nor even one of mathematical validity: it depends upon the actual structure of the world. Which borders on the next condition.

(vii) *Factual interpretation should maximize truth.* The semantic assumptions of a scientific theory should contribute to producing a maximally true theory. As with the previous conditions, this one is easier legislated than lived up to. Since factual truth depends on both the mathematical formalism and the semantic assumptions, the goal of maximal truth can only be attained by a mutual adjustment of these two components. The test of correctness of semantic assumptions is of course the truth of the theory as a whole. But we can never check the entire theory and we should not expect it to be completely true. Hence even a strong confirmation of the theory provides no final assurance that the semantic assumptions are right. And if the tests are unfavorable then we may blame either the formalism or the semantic assumptions and attempt to reform either. Whatever the outcome of the tests we cannot be sure about the adequacy of the interpretation. We must take risks and be prepared to lose. In sum, interpretation is just as tentative as formalism – and both are prior to tests. By the same token interpretations can be changed in the interest of truth. If a theory fails to pass some tests for truth it need not be rejected *in toto*: some of it may be salvaged by partially modifying its formalism or its interpretation or both. In any case interpretation is prior to truth valuation and should maximize truth value.

This last condition leads us to the next point – the confusion between interpreting and stipulating truth conditions.

3.7. *Factual Interpretation and Truth*

Interpretation maps constructs either into further constructs (the case of μ) or into facts (the case of ϕ). In either case interpretation is prior to truth valuation: the latter depends upon the former. Thus consider the abstract formula ⌜For every x and z there is at least one y such that $x \circ y = z$⌝. Unless we interpret the individual variables and the operation we cannot even ask whether the formula holds. A formula in abstract mathematics holds or fails to hold *under some interpretation* (or in some model) or other. Of course we are chiefly interested in interpretations conducive to truth, so that an interpretation that fails to satisfy this condition will be given up. Likewise in factual science interpretation comes before truth valuation even though an unfavorable outcome of the latter may force us to reinterpret the given mathematical formalism. In sum, both in mathematics and in factual science *only interpreted formulas can be tested for truth*, and only such tests allow us to assign truth values. In a nutshell, the process looks like this:

$$\textit{Formulation} \rightarrow \textit{Interpretation} \rightarrow \textit{Testing} \rightarrow \textit{Truth valuation}.$$

The same can be seen from another angle: interpretation and truth valuation are utterly different functions. Let us confine ourselves to factual interpretation ϕ and to the assignment $\mathscr{V}$ of degrees of truth of fact. (But similar considerations hold for mathematical interpretation and assignment of formal truth.) For one thing ϕ applies not only to full formulas but also to their nonlogical constituents, whereas $\mathscr{V}$ is a (partial) function on statements. For another, whereas the range of ϕ is a factual domain, the range of $\mathscr{V}$ is a set of truth values, e.g., 0 and 1. Consequently *to give the semantics of a scientific theory does not involve giving truth conditions*, let alone truth values: it only requires *specifying the interpretation map* ϕ.

However, according to a widespread view interpreting involves or even consists in giving truth conditions, perhaps even truth values. Thus Carnap: "By a *semantical system* (or interpreted system) we understand a system of rules, formulated in a metalanguage and referring to an object language, of such a kind that the rules determine a *truth-condition* for every sentence of the object language, i.e. a sufficient and necessary condition for its truth. In this way the sentences are *interpreted* by the rules, i.e. made understandable, because to understand a sentence, to know what is asserted by it, is the same as to know under what conditions it would be true. To formulate it in still another way: the rules determine the *meaning* or *sense* of the sentence" (Carnap, 1942, p. 23; see also p. 203). And, a quarter century later, Davidson (1967, p. 310): "to give truth conditions is a way of giving the meaning of a sentence".

This influential view is a version of the verification doctrine of meaning suggested by Frege and proposed by the operationists, Wittgenstein, and the Vienna Circle. It is so confusing that the reasons for rejecting it bear some hammering. First, although the doctrine looks plausible for the propositional calculus, where the sense of the connectives may be said to be given by their truth tables, the view fails for predicate logic. Here both the individual variables and the predicate variables have to be interpreted in a manner that is independent of truth, as shown in any standard logic textbook (e.g. Mendelson, 1963, pp. 49ff.; Shoenfield, 1967, pp. 61ff.; Suppes, 1957, pp. 64ff.). Second, before we set out to find out the truth value of a formula we must know what it "says" about what: fancy placing truth conditions on an uninterpreted formula. Third, truth depends upon interpretation, not the other way around. Thus $\ulcorner(\exists x)\, Gx\urcorner$ is true for $\phi(G)=$Gipsy, and false for (or "under") $\phi(G)=$Ghost. Fourth, except for idealist philosophers the assignment of degrees of factual truth is not a matter of semantics but of observation and scientific inference. Semantics cannot even devise truth conditions for scientific hypotheses and theories: this is a matter for methodology. Thus consider a theoretical statement of the form

$$t=\ulcorner P(s,u)=n\urcorner$$

evaluated in the light of a piece of empirical evidence in the form

$$e=\ulcorner\text{Average of measured values of } P(s,u)=n'\pm\varepsilon\urcorner,$$

where P is some property of a system s, n is the calculated value and n' the measured value (both in units u), while ε is the experimental error. Then a "truth condition" for t that is universally agreed on (without the assistance of the available semantical theories) is this:

t is true relative to e, to within ε, *iff* $|n-n'|\leqslant\varepsilon$.

The actual value of the experimental error ε will depend on the state of the experimental art: it is no business of semantics. (For the empirical assessment of truth values see Bunge, 1963a, pp. 127ff; and Bunge, 1967a, II, pp. 301ff.)

In short, interpretation and truth are related but not the way operationist semantics has it. Truth depends upon interpretation, which in turn should be subject to revision in the light of the tests for truth. A formula will hold or fail to hold (exactly or approximately), under a given interpretation, whereas alternative interpretations may render the formula meaningless or utterly false. This holds for mathematics as well as for science. So much for one of the worst muddles in the history of philosophy.

3.8. *Interpretation and Exactification*

Factual constructs come with different degrees of exactness and clarity. The most exact and clear of all are those belonging to a theory, i.e., the theoretical factual constructs. According to Def. 3 (Sec. 3.2) a construct c of this kind is a mathematical construct m together with a factual interpretation ϕ, i.e., $c = \langle m, \phi \rangle$. Consequently to *exactify* a factual construct consists in either revealing or assigning its formal component m. And to *elucidate* a factual construct consists in either revealing or assigning its semantic component ϕ. If we disclose the form or the content of a construct we engage in analysis; if we assign either of them we build or reconstruct a fragment of one of the theories housing the construct of interest.

In principle every *bona fide* scientific construct can be both exactified and elucidated, namely by either incorporating or expanding it into a theory or, if it belongs already to some theory, by analyzing or reconstructing the latter. Even initially obscure concepts can be exactified and elucidated. A good example is the one of disposition, tendency, propensity, or bent, which pervades both factual science and metaphysics. This intuitive notion can be analyzed into two distinct concepts: those of *causal propensity* and *chance propensity* (cf. Bunge, 1974b). An instance of the former is solubility: dissolution is an outcome of the mixing of the soluble substance with a proper solvent under suitable circumstances. Whenever these conditions are met dissolution occurs. Not so chance propensity, as exemplified by the emission of light by an atom or the learning of an item by an animal: even if the necessary conditions are met, the event occurs only with a certain probability – i.e., there seem to be no conditions both necessary and sufficient for the event to happen. Let us focus on this second concept of tendency, which is by far the more baffling and probably the most fundamental of the two.

The intuitive or pretheoretical concept of chance propensity is exactified by means of the mathematical concept of probability. And any specific concept of chance propensity is elucidated by incorporating it into a factual theory; for example, every concept of learning disposition or ability is elucidated by the corresponding stochastic learning theory. Exactification, though essential, is not sufficient to turn a specific concept of chance propensity into a semantically precise concept, because the mathematical theory of probability is not committed to any particular factual interpretation. We must specify also the interpretations of the arguments and values of the probability function. (Recall Example 1 in Sec. 3.4.) This may be done in the following way. Let the pretheoretical idea be that of the tendency or ability of a system σ of kind Σ to make a transition from an initial state A to a final state B. (For example, Σ could be a strain of albino rats, A the state of ignorance concerning the proper way to run a T maze, and B some stage in the learning process.) The explicans of that relatively obscure notion of ability is the ordered pair $\langle Pr(B \mid A), \phi \rangle$, where $Pr(B \mid A)$ is the conditional probability of B given A, and ϕ the interpretation map defined by the following value assignments:

$$\phi(\sigma) = \text{System of kind } \Sigma \tag{1}$$
$$\phi(A) = \text{Initial state of } \sigma \tag{2}$$
$$\phi(B) = \text{Final state of } \sigma \tag{3}$$
$$\phi(Pr(B \mid A)) = \text{Strength of the propensity for } \sigma \text{ to jump from } A \text{ to } B. \tag{4}$$

In other words, the coarse or presystematic idea of the tendency for σ to go from A to B is given a refined (exact and lucid) expression by the theoretical factual construct $\langle Pr(B \mid A), \phi \rangle$, which belongs to some theory concerning certain traits of the systems of kind Σ – a theory whose mathematical formalism includes some fragment of the mathematical theory of probability. While the latter is in charge of the *exactification* of the concept of chance propensity, the interpretation assumptions (1) to (4) provide an *elucidation* (or semantic clarification) of it. Let us insist that the interpretation assumptions are not part of the exactification procedure but external to it. If we were to regard them as belonging to the exactification process we would fall into a circle: we would be explaining propensity as propensity.

The preceding considerations solve one of the problems posed by the so-called propensity interpretation of probability championed by Popper (1959). The problem is to answer the charge that nothing is gained and much is lost by interpreting the clear concept of probability in terms of the obscure concept of propensity. Our answer is this: There is nothing wrong with adopting at the same time the propensity *interpretation* of probability, i.e.,

$$\phi\,(\text{Probability}) = \text{Propensity},$$

and the probability *exactification* of propensity, i.e.,

$$\varepsilon\,(\text{Propensity}) = \text{Probability},$$

as long as the two are not mixed up. Whereas interpretation assigns a factual content to a definite mathematical construct, exactification transforms an inexact construct into a definite mathematical object. Shorter: whereas exactification bears on presystematic concepts, interpretation enriches exact concepts.

Moreover just as the probability exactification of chance propensity is consistent with the propensity interpretation of probability, both are compatible with the *frequency estimate* (evaluation) of probability values. For example, in game theory the probability that a player chooses a given strategy may be interpreted as his propensity to adopt such a choice (cf. Rapoport, 1966); and this value can be estimated by observing the actual frequency of such an event. What is not possible is to give a *frequency interpretation* of probability. For one thing, probability and frequency are different functions: the latter is defined, for every sampling procedure, on a finite subset of the whole probability space; and the range of the frequency function is not the whole real interval [0, 1] but the collection of fractions in it. (Cf. Bunge, 1969.) Consequently if probabilities were *interpreted* as frequencies, the typical theorems of the calculus of probability, such as the laws of large numbers, could not even be stated, as they concern precisely *differences* between probabilities and frequencies. (For further criticisms of the frequency theories of probability see Fréchet (1939) and Suppes (1967, Ch. 3.)) Interpretation is a strictly conceptual operation not to be confused with numerical estimation, in particular measurement. Such a confusion amounts to mistaking semantics for pragmatics – something that deserves a separate section.

4. Pragmatic Aspects

4.1. *Pragmatic Interpretation*

School teachers find it effective to clarify mathematical and scientific ideas with reference to human operations. Thus ⌜3+2=5⌝ can be made clear and plausible by fingering, and Archimedes' law of the lever can be literally felt when riding a see-saw. These are examples of *pragmatic interpretation* or interpretation in terms of human actions. In Sec. 2 we mentioned some pragmatic interpretations of the propositional calculus. Table 6.4 exhibits pragmatic interpretations of a few typical formulas of

TABLE 6.4

Examples of pragmatic interpretation of typical formulas

Construct	Semantic interpretation	Pragmatic interpretation
Pa	The individual a has the property P.	Someone has proved or observed that a is a P.
$(\exists x)_U Px$	There is at least one object in U with the property P.	At least one object in the observed collection $T \subseteq U$ has been found or may be found to be a P.
$(x)_U Px$	All objects in U have the property P.	Every object in the observed collection $T \subseteq U$ has been found or may be found to be a P.
$A \vdash B$	A entails B.	B is provable from A.
$f(x)=y$	The f-ness of x equals y.	The result of determining (computing or measuring) f at x is (nearly) y.
$P_n(x)=0$	The nth degree polynomial P_n in x equals 0.	Find the values of x that annul the nth degree polynomial P_n in x.

the predicate calculus. The pattern is this: every construct in a set C is assigned one item in a set H of human actions. In short, $\pi: C \to H$. In other words, π is a rule or instruction for handling a construct with a definite means and a definite goal.

There can be no quarrel with pragmatic interpretations when employed as didactic crutches – particularly if being reassured that the crutches will be dropped in due time. Nor should it be objected to translating formulas into instructions or commands for purposes of computer processing, lab checking, or action – particularly if the formulas are allowed to retain a content of their own independent of the way they are used or tested. What is inconvenient is to have to walk on crutches all life long; even worse, to be under the delusion of being a computer or of being chained to some measuring device. In other words, what is objectionable is to mistake a construct for a pragmatic interpretation of it. It is even worse to dignify this confusion with the name of a philosophy – say operationism, operative logic, or mathematical intuitionism. In short, while pragmatic interpretations are occasionally valid and useful (though always restricted to a minute subset of the collection of constructs), pragmatist semantics is untenable.

Most pragmatic interpretations are *adventitious* in the sense of Sec. 3.6. Indeed, in most cases they do not match the structure of the formula they bear on, as they refer to individuals (e.g., observers) and actions (e.g., measurements) not represented in the formula by any variables. For example, the orthodox interpretation of an eigenvalue α_k of a quantum-mechanical operator A_{op} reads: "α_k is a possible result of measuring the property represented by A_{op}". This interpretation is adventitious because neither A_{op} nor α_k (nor the corresponding eigenfunction) contain any variables capable of representing the measuring device (which one?) or the experimenter (who?). (Cf. Ch. 3, Sec. 4.3.)

If we were to outlaw all adventitious interpretations, few pragmatic interpretations would remain. As long as we are aware of the narrowness and arbitrariness that adventitious interpretation may land us on, we may afford to adopt a wider concept of interpretation validity. We propose the following conditions for regarding a pragmatic interpretation of a formula as *valid* (Bunge, 1969):

(i) There should exist a scientific theory containing the formula and assigning to it a semantic (mathematical or factual) interpretation. In other words, the formula to be interpreted (*a*) must be available to begin with and (*b*) must have a fairly definite content independent of the many ways it can be manipulated. (Imagine rushing to read a new scientific theory in operational terms before finding out what the sense and the reference of the theory are.)

(ii) Sufficient information, theoretical and empirical, must be at hand to justify as well as to carry out the operations called for or described by the pragmatic interpretation. If the construct on which the interpretation bears happens to represent an unobservable entity or property, as is so often the case with science, then additional hypotheses or theories linking the unobservable to observable items will be needed. (I.e., objectifiers or indicators will be needed, and these will usually rope in further theories.) Otherwise the proposing of a pragmatic interpretation would be a game like casting horoscopes or interpreting dreams. In other words valid pragmatic interpretation, even when adventitious, is a matter of law not convention: there should exist a lawful relation between the referent of the construct and the human action the rule prescribes. In sum, pragmatic interpretation should be grounded.

Pragmatic interpretation occurs in experimental science and in technology. The experimentalist may well read $\ulcorner y=f(x) \urcorner$ as "To infer y measure x" provided f is defined and the semantic interpretation tells him what these symbols stand for. Likewise the engineer may read the same formula as, say, "To get output y apply input x" – as long as the underlying theory supplies him with the sense and reference of the formula and provided experiment encourages him in taking the assumed functional relation as close enough to the truth. Pragmatic interpretations such as these are valid albeit adventitious: they rely on the formula concerned as well as on its semantic interpretation. Similarly with every other pragmatic interpretation: if valid it will rest on a previously assigned semantic interpretation. First get to know, then apply your knowledge.

There is no room for pragmatic interpretation in scientific theories. A theoretical formula refers to some concrete system (cell, society, or what have you) not to the way that very formula is to be tested or applied. Even the sciences of action, such as operations research and political science, regard their referents as objects. Consequently their formulas are first assigned a semantic interpretation, then they may be put to use as rules of procedure. We had to emphasize the dependence of pragmatic interpretation upon semantic interpretation because of the strong human tendency, called *anthropomorphism*, to read everything in terms of human feelings and actions. We must rid semantics of any association with such a tendency if we want it to account for the objectivity of science.

4.2. *The Interpretation Process*

Interpretations do not come out of the blue and, once proposed, they need not stay. Viewed historically, interpretation is a process. In some cases the formalism of a theory and its interpretation evolve hand in hand. In others the embryo is an intuitive idea in search of a formalism: this may have been the case with Newtonian mechanics, Maxwellian electrodynamics, and Einsteinian gravitation theory. Finally the converse process, namely the building of a formalism in search of an interpretation, can also happen: as a matter of fact this seems to have been the case, to a large extent, with quantum mechanics (Dirac, 1942; Heisenberg, 1955). Consequently there are no hard and fast rules for "discovering" the semantic assumptions of a scientific theory: some investigators proceed in one way, others differently. It is up to the psychology of science, not to semantics or even methodology, to discover what makes researchers tick, in particular what makes them guess that a given formula should be interpreted in a certain way.

Moreover no interpretation is likely to be final. Every theory in a process of growth suffers both mathematical and semantical adjustments. Even classical theories are still undergoing changes of both kinds (cf. Truesdell and Toupin, 1960). In particular the new interpretation may be at variance with the original intentions of the first theorist. The psychologist and the historian of science may wish to ask him what interpretation he had in mind but won't be able to discover all the possible unintended interpretations – if only because most of them will never occur to anyone. In any case the concept of intention, that occurs in the usual phrase 'intended interpretation', is a psychological one and therefore beyond the reach of semantics. Whether a particular outcome, semantic or other, was originally intended, is a psychological and historical problem. Therefore it is misleading to define interpretation as an *intended model* of a formalized language (Kemeny, 1956). For the same reason it is unsatisfactory to mention the *intended interpretation* of a formalism without stating it explicitly; and, once so stated, it is no longer intended. The semantic assumptions of a theory, whether they are the originally intended (or standard) ones or not, should be formulated as explicitly as the remaining assumptions if we are to have objectivity and the possibility of rational argument.

The need for arguing about interpretation matters is not always felt. It seems most acute in highly developed fields – but there it is often repressed. Every theoretical biologist knows that it is far easier to interpret the solution to a problem in mathematical biology than it is to formulate the problem. The opposite situation is the rule in physics, where it is far easier to formulate a problem and even to perform the computational tasks it calls for, than to find an adequate interpretation of the solution. Why this difference? In biology there are hardly any comprehensive theories supplying a general framework for the formulation of problems. Except in choice areas such as biophysics and genetics, nearly every problem has got to be treated separately, often leaning more heavily on physics or chemistry than on biology. Usually theories have to be built from scratch, sometimes inaugurating whole new branches of biology in the process. As a compensation the goal is more modest: the variables involved are fewer, they are often better understood and are frequently linked in simpler ways than the variables occurring in theoretical physics and chemistry. We can expect however that, as biology grows in depth, it will pose as many and as hard interpretation problems as physical theories do at present.

Regretfully, rational argument about the semantic assumptions of scientific theories is sometimes discouraged or even hushed up. Even complex theories like quantum mechanics and quantum electrodynamics, ridden as they are with unsettled interpretation problems, are often regarded as unproblematic and all discussion about their semantic assumption as a waste of time (Rosenfeld, 1961). There are several possible motives but not a single reason for adopting such a dogmatic and unhistoric stand. One is the longing for certainty. Another is the belief that foundations problems are settled by popular philosophical discourse rather than by digging out the axiomatic foundations of the theory concerned. A third possible cause is a faulty semantics of science, one holding that all that really counts in a scientific theory is its mathematical formalism. If this were true, then the producing of any new formula or any new set of numbers would be a valuable contribution to scientific knowledge, whereas the proposing of a more cogent interpretation of a theory would be insignificant. This attitude is widespread among scientists who must devote most of their time to solving difficult computation problems, e.g., with the help of perturbation theory. They take the basic equations for granted and consider themselves lucky if once in a blue moon they can find solutions in closed form – which are the ones best suited to interpretation. Since they have little time left to ponder over the interpretation of their very starting points, they have no patience with anyone's telling them that interpretation is always problematic, hence deserving of closer scrutiny. But of course the belief is mistaken. Since a theory in science is a formalism together with an interpretation, a change in the latter produces a new theory. Besides, some interpretations deserve being reformed because they are wrong. Hence disputes over interpretation matters are as important as arguments over mathematical points. What is true, and unfortunate, is that the standards of argumentation over semantical problems are so much lower than the standards of mathematical argument. It behooves the philosopher to upgrade those standards by building a semantical theory competent to deal with live science.

5. Concluding remarks

In our view, since meaning is sense together with reference, a meaning assignment is an assignment of both sense and reference. Such an assignment involves an interpretation of the symbols concerned and eventually also an interpretation of the constructs designated by the symbols – as when the letter '*N*' is first read as the cardinality of a set, then as the population of a group of organisms. However, we do not regard interpretation as meaning assignment. One reason for not identifying these two concepts is that, while interpretation may bear either on signs (e.g., predicate letters) or on constructs (e.g., functions), we construe meaning as a property of constructs only (see Ch. 7). Another reason is that not every interpretation assigns significance: some interpretations result in nonsignificant expressions. For example, if the predicate letter *P* in '5 is *P*' is interpreted as 'painful', a nonsignificant sentence results. Interpretation, though necessary, is insufficient to guarantee significance. Significance derives from meaning, which is in turn a conceptual matter.

Even assuming that we grasp the general concept of meaning we may not know how to go about assigning or finding out specific meanings. Placing the given construct (concept or proposition) in a given context (e.g., a theory) is surely necessary to that end since meaning is contextual, but may not suffice. Thus the axioms of an abstract theory, such as Boolean algebra, determine the (mathematical) sense of the theory but, fortunately, fail to specify any of the possible referents of the theory. In other words, the sets involved in the abstract theories are abstract: they consist of faceless individuals. It is only when adjoining an interpretation that definite individuals are characterized by the theory. And only semantic assumptions of the factual type indicate that such individuals are factual items. The semantic assumptions of a factual theory determine, *inter alia*, the reference class of the theory and thus contribute to sharpening its factual sense. In terms of the symbolism or language of a scientific theory: the axioms of such a theory (all the axioms) determine both the sense and the reference the symbolism points to. Or, as we shall say, they jointly determine the meaning of the factual theory. This, in a nutshell, is the theory of meaning to be developed in the next chapter.

CHAPTER 7

MEANING

We are now in a position to face the chief scandal of semantics. We shall start by distinguishing the significance of a sign from the meaning of the construct it designates. We shall then proceed to formulate and discuss our view that meaning is no more and no less than sense together with reference. If either component changes then a changed meaning, i.e., a different construct ensues. Since our previous investigations have shown us how to disclose sense and spot referents, we shall be in a position to compare meanings. In particular we shall be able to ascertain whether two given constructs have the same meanings so that their respective symbols will be synonymous. Finally we shall discuss some of the difficulties that have hampered the clarification of the meaning concept.

1. Babel

Although the concept of meaning has been the subject of active inquiry since Socrates, and at the center of analytic philosophy for half a century, it is still far from clear. To be sure there have been heaps of brilliant meaning analyses and plenty of talk about the theory of meaning and the theory of reference, particularly about the virtues such theories ought to exhibit. But in fact no theory proper has been offered that does justice to the two aspects of meaning traditionally discerned, namely sense (or connotation) and reference (or denotation). And none of the existing theories of meaning, not even those proposed by philosophers of science, has been of any help whatever in performing a semantic analysis of a piece of live science or in teaching scientists to speak sensibly about the meaning of their own creations. No wonder many physicists still claim that the meaning (not just the worth) of a theoretical item is determined by observation procedures. No wonder chemists are fond of saying that each triplet of bases means (not just specifies or determines) a particular aminoacid. No wonder geneticists sometimes hold that mutations can produce meaningless sequences – rather than biologically dysfunctional proteins. In sum half a century of talk about meaning has been useless to scientists and has, if anything, increased confusion among philosophers. It has resulted in a Babel.

The views on meaning come in several degrees of formal sophistication but their gist can be conveyed in plain words. The most influential among the contemporary views are displayed schematically in the following list.

1 *Psychologism:* meaning is either thought or intention or understanding.
2 *Pragmatism:* meaning is usage.
3 *Operationism:* meaning is operation (computation or measurement).
4 *Verificationism:* meaning is truth condition.
5 *Epistemic view:* meaning is information.
6 *Nihilist view:* there are no meanings.
7 *Referential view:* meaning is the thing referred to.
8 *Intensional view:* meaning is either intension or content.
9 *Dualist view:* meaning has two dimensions – intension and extension.
10 *Synthetic view:* meaning is composed of sense and reference.

The first six views are characteristically modern. Psychologism is the oldest of them but was not made academically respectable until Brentano expounded it one century ago. It was espoused at one time by Russell (1919b). The second, linguistic pragmatism (and in particular dictionarism), was propounded by the later Wittgenstein. The third, operationism, may be traced back to Peirce and Dingler (1907) and has been in the air natural scientists have been breathing ever since. The fourth, verificationism, may be traced back to Frege and was one of the battle cries of the Vienna Circle. The fifth, informationism, was born in the 1950's. And the sixth, nihilism, is a cry of despair over the failure of all the previous opinions. We have criticized and rejected all six in Chs. 2, 4 and 5.

The remaining four views have much longer roots and are far sounder. The referential doctrine goes back to the medieval nominalists, particularly Ockham and Buridan, and their modern heirs, especially Hobbes. The intensionalist doctrine has been a constant component of idealism probably since Plato and was particularly vivid in Leibniz and Bolzano (1837). The dualist doctrine was sketched in the Port-Royal *Logique*

(1662) and was later revived, but also obscured, first by Frege (1891, 1892), then by Lewis (1944, 1951).

(We may as well list the weaknesses of Frege's influential doctrine of meaning in order to forestall any confusion between it and the next or synthetic view. To begin with, the great Frege did not employ a consistent terminology: for example, he often exchanged *Bedeutung* (our "reference") and *Bezeichnung* (designation). He did not distinguish clearly between reference and extension. He often construed *Sinn* (sense) in a psychologistic fashion, namely as the thought expressed by a sentence. He identified the *Bedeutung* of a statement with its truth value and ominously assigned truth conditions the job of determining senses. Finally Frege had no semantic theory to speak of: he never advanced beyond some unsystematic remarks – albeit often illuminating and always provocative. Frege's importance for semantics seems to lie in that (*a*) he stressed a number of distinctions – notably between symbol and construct, concept and extension, and sense and truth, and (*b*) he called the attention of others, notably Russell and Carnap, to semantic problems. For a very different evaluation see Dummett (1973, Ch. 19).)

Finally the tenth view – the synthetic doctrine – seems to have a medieval pedigree too. It was rescued from oblivion and popularized by J. S. Mill: his *System of Logic* (1843) was so influential that his distinction between the connotation and the denotation of a term has been incorporated into common speech. This view appealed to common sense and, since Mill was a champion of positivism, his revindication of connotation or sense, against the grain of nominalist referentialism, was above the suspicion of Platonism. Our own view is but an elaboration of Mill's and of Williams' (1937), which incorporated Mill's without its positivism. We have called it *synthetic* for the following reasons: (i) it applies not only to terms and other linguistic expressions but also (and primarily) to their conceptual designata; (ii) while it distinguishes sense from reference, it combines them into a single idea with a definite mathematical status – namely the sense-reference ordered pair; (iii) far from being a stray view it is an outcome of our theories of sense and of reference expounded in previous chapters. Let us peek at it before going into details.

To begin with we stipulate the kinds of object that can possess meaning. They are certain symbols and all constructs. In order to ward off confusion we shall call these two possibilities by different names: we shall say that (some) *symbols signify* and that (all) *constructs mean.* Furthermore we shall take meaning as primary and shall define it by its two components: sense and reference. And we shall construe signification, a property of certain signs, as the composition of designation and meaning. Graphically:

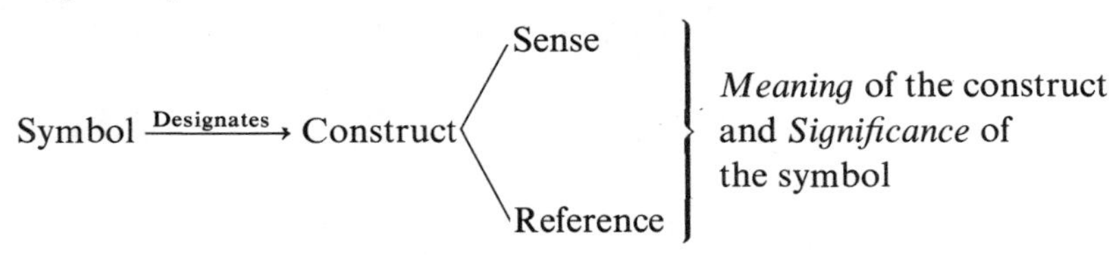

For example we shall say that the term 'man' designates the (or rather *a*) concept of man, the sense of which is given by anthropology, its reference class being the whole bunch of humans. These two constitute also the significance or vicarious meaning of the word 'man'. On the other hand any sign that fails to designate a construct will be assigned an empty significance. For example a punctuation mark stands for no construct and is therefore devoid of significance. In sum if a sign is significant it is so vicariously, namely via some construct. This construal of 'meaning' avoids both nominalism and the variety of hylemorphism consisting in endowing mere marks with semantical properties. And it does not commit us to Platonism, for we do not adopt the ontological hypothesis that constructs have an independent being. Furthermore our view formalizes the opinion of those linguists who hold that a word has two semantic functions: one is to denote, the other is to encapsulate a whole system of generalizations and associations (Luria, 1961).

2. The synthetic view

2.1. *Meaning as Sense cum Reference*

In the previous section we agreed to regard meaning as a property of constructs (concepts, propositions, or theories). Now we propose to analyze the meaning of a construct as its sense together with its reference, as depicted in Figure 7.1 and exemplified in Table 7.1.

The stipulation that sense and reference are to be taken as the two components of meaning is literal not metaphorical. Calling $\mathcal{S}(c)$ the sense and $\mathcal{R}(c)$ the reference class of a construct c, our proposal can be symbolized as

$$\mathscr{M}(c)=\langle \mathscr{S}(c), \mathscr{R}(c)\rangle.$$

Now, as we saw in Chapters 4 and 5, $\mathscr{S}$ maps constructs into sets of constructs, i.e. $\mathscr{S}: C \to \mathscr{P}(C)$. And, according to Chapter 2, $\mathscr{R}$ maps constructs into sets of objects of any kind, i.e. $\mathscr{R}: C \to \mathscr{P}(\Omega)$, where $\mathscr{P}(\Omega)$ is

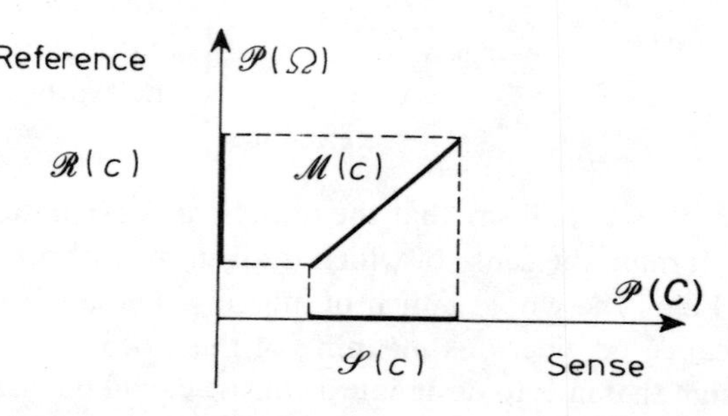

Fig. 7.1. Sense and reference as the components of meaning.

TABLE 7.1
The two components of meaning of the rat race theory in Chap. 6, Sec. 3.2

Construct	Meaning components	
	Sense	Reference
S	Set	Rats
Δ	Joining a rat race	Rats

the set of all subsets of the collection Ω of objects. The pairs $\langle \mathscr{S}(c), \mathscr{R}(c)\rangle$ define uniquely a third function

$$\mathscr{M}: C \to \mathscr{P}(C) \times \mathscr{P}(\Omega)$$

such that

$$p \circ \mathscr{M} = \mathscr{S} = \text{The first projection } (= \text{component}) \text{ of } \mathscr{M}$$
$$q \circ \mathscr{M} = \mathscr{R} = \text{The second projection } (= \text{component}) \text{ of } \mathscr{M}$$

That is, the two triangles in the following diagram commute:

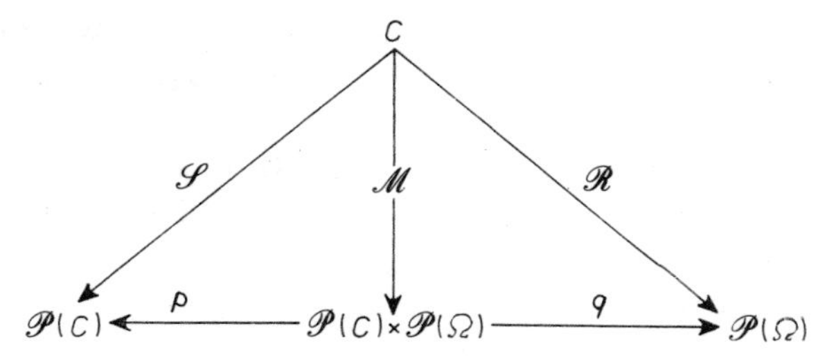

In other words, we lay down the following

DEFINITION 7.1 Let Ω be the universe of objects and $C \subset \Omega$ the collection of constructs. Call $\mathscr{S}: C \to \mathscr{P}(C)$ the (full) sense and $\mathscr{R}: C \to \mathscr{P}(\Omega)$ the reference functions. Then the bijection $\mathscr{M}: C \to \mathscr{P}(C) \times \mathscr{P}(\Omega)$, such that $\mathscr{M}(c) = \langle \mathscr{S}(c), \mathscr{R}(c) \rangle$ for c in C, is called the *meaning function*, and its value $\mathscr{M}(c)$ at c is called the *meaning of c*.

Remark 1 Note that the full sense is involved in the preceding – i.e., the union of the purport and the import of the construct concerned (recall Ch. 5, Sec. 5). *Remark 2* This construal of meaning does not depent on the concept of truth, as both sense and reference are prior to any assignment of truth value. Both Lewis' (Lewis, 1944, 1951) and the author's (Bunge, 1967a) previous definitions as the intension-extension pair are unacceptable for this reason. (They are not identical because Lewis construes intensions in a referential fashion.) *Remark 3* On any extensionalist semantics theology is meaningful or meaningless depending on one's religious beliefs. According to our view theological statements can be perfectly meaningful in their own contexts, which determine both their sense and their reference. Belief and disbelief must rest on the assignment of extensions, not of sense or of reference. Thus while for a theist $\mathscr{E}$(Creator) $= \mathscr{R}$(Creator) $=$ {God}, for an atheist $\mathscr{R}$(Creator) $=$ $=$ {God}, but $\mathscr{E}$(Creator) $= \emptyset$. Consequently if anyone wished to argue for or against a particular religion he could not count on our semantics: he would have to resort to alternative means. In particular the disbeliever won't get away with the cheap assertion that theology is nonsense. (But he may succeed in showing that some theologies are contradictory, or that all of them lack positive empirical evidence.) On the other hand the claim that existentialism and Zen Buddhism make hardly any sense stands.

We can now give an exact answer to a question left over from the days of logical positivism, viz., Are tautologies meaningful? In Ch. 2, Sec. 3.3 we saw that a tautological construct refers to anything; if universal, like $\ulcorner(x)(Px \vee \neg Px)\urcorner$, it refers to the totality Ω of objects. And in Ch. 5, Sec. 4, we saw that the sense of a tautological construct in a context $\mathbb{C} = \langle S, \mathbb{P}, D\rangle$ with underlying logic L equals $S \cup L$. Consequently if t is any universal tautology in L,

$$\mathcal{M}_{\mathbb{C}}(t) = \langle S \cup L, \Omega\rangle.$$

If the context in question is strictly logical, i.e. if $S = L$, then t "says" whatever L "says", and this about any objects:

$$\mathcal{M}_L(t) = \langle L, \Omega\rangle.$$

But of course the extralogical meaning of t in L is $\langle \emptyset, \Omega\rangle$. I.e. within logic the tautologies "say" nothing extralogical about everything. And when associated with some extralogical body of knowledge they "say" everything the latter "says" because they get stuck to every bit of it. Hence logic cannot, by itself, teach us anything about the world even when we make it speak about the world: logic is not ontology. Whatever logic succeeds in teaching us about the world it does so by associating with extralogical contexts. In conclusion tautologies are meaningful even if they do not inform us about the world. (More in Bunge, 1974a.)

Consider now the whole meaning space, i.e. the totality of values of the meaning function $\mathcal{M}$. Take the set $\mathbb{P}$ of all predicates (alternatively propositions) concerning a fixed universe of discourse $D \subset \Omega$. Further, call $\mathcal{M}(\mathbb{P})$ the totality of meanings carried by the constructs in $\mathbb{P}$. The following operations can be defined in $\mathcal{M}(\mathbb{P})$: for any p and q in $\mathbb{P}$,

Meaning addition: $\mathcal{M}(p) + \mathcal{M}(q) = \langle \mathcal{S}(p) \cup \mathcal{S}(q), \mathcal{R}(p) \cup \mathcal{R}(q)\rangle$,

Meaning product: $\mathcal{M}(p) \dot{\times} \mathcal{M}(q) = \langle \mathcal{S}(p) \cap \mathcal{S}(q), \mathcal{R}(p) \cap \mathcal{R}(q)\rangle$,

Meaning complement: $-\mathcal{M}(p) = \langle \overline{\mathcal{S}(p)}, \overline{\mathcal{R}(p)}\rangle$.

Clearly the two binary operations are associative and commutative. Moreover meanings are idempotent: $\mathcal{M}(p) + \mathcal{M}(p) = \mathcal{M}(p)$ and similarly for the product. Also, multiplication distributes on both sides over addition:

$$\mathcal{M}(p) \dot{\times} (\mathcal{M}(q) + \mathcal{M}(r)) = (\mathcal{M}(p) \dot{\times} \mathcal{M}(q)) + (\mathcal{M}(p) \dot{\times} \mathcal{M}(r))$$
$$(\mathcal{M}(p) + \mathcal{M}(q)) \dot{\times} \mathcal{M}(r) = (\mathcal{M}(p) \dot{\times} \mathcal{M}(r)) + (\mathcal{M}(q) \dot{\times} \mathcal{M}(r)).$$

Finally, by combining complement and product we build the null or least element $\square$ of $\mathscr{M}(\mathbb{P})$, whilst combining complement and addition yields the unit or last element $\blacksquare$:

Null meaning: $\mathscr{M}(p)\dot{\times}(-\mathscr{M}(p))=\langle\mathscr{S}(p)\cap\overline{\mathscr{S}(p)},\ \mathscr{R}(p)\cap\overline{\mathscr{R}(p)}\rangle=$
$=\langle\emptyset,\emptyset\rangle=\square.$

Universal meaning: $\mathscr{M}(p)+(-\mathscr{M}(p))=\langle\mathscr{S}(p)\cup\overline{\mathscr{S}(p)},\ \mathscr{R}(p)\cup$
$\overline{\mathscr{R}(p)}\rangle=\langle\mathscr{P}(\mathbb{P}),\mathscr{P}(D)\rangle=\blacksquare.$

The set $\mathscr{M}(\mathbb{P})$ is not closed under addition and multiplication because, as we saw in Ch. 5, in general $\mathscr{S}(p)\cup\mathscr{S}(q)$ does not equal the sense of some compound of p and q. Only intensions behave this way. Hence if we restrict sense to intension the preceding considerations prove that the structure $\langle\mathscr{M}(\mathbb{P}),\square,\blacksquare,+,\dot{\times},-\rangle$ is a ring of idempotents with unit and zero, i.e. a Boolean ring. We shall not pursue this line but shall instead analyze two notions of meaning relation.

DEFINITION 7.2 Let $p, q\in\mathbb{P}$ be either predicates or propositions with definite meanings. Then the meaning of p is *part* of ($\leqslant$) the meaning of q iff the meaning of p does not add anything to that of q:

$$p\leqslant q=_{df}\mathscr{M}(p)+\mathscr{M}(q)=\mathscr{M}(q).$$

DEFINITION 7.3 Let $p, q\in\mathbb{P}$ be either predicates or propositions with definite meanings. Then p and q are *semantically unrelated* ($\urcorner\!\llcorner$) iff the product of their meanings is nil:

$$p\urcorner\!\llcorner q=_{df}\mathscr{M}(p)\dot{\times}\mathscr{M}(q)=\square.$$

COROLLARY 7.1 Let $p, q\in\mathbb{P}$. Then

(i) If $p\leqslant q$ then $\mathscr{S}(p)\subseteq\mathscr{S}(q)$ and $\mathscr{R}(p)\subseteq\mathscr{R}(q)$.

(ii) If $p\urcorner\!\llcorner q$ then $\mathscr{S}(p)\cap\mathscr{S}(q)=\emptyset$ and $\mathscr{R}(p)\cap\mathscr{R}(q)=\emptyset$.

Finally consider the whole lot $\mathscr{T}$ of theories with a common meaning core – e.g. the theories in mathematical linguistics. Because of the shared meaning the following sets will be nonempty for any two members T_i, T_k of $\mathscr{T}$:

$$T_i\sqcap T_k=\{s\mid T_i\vdash s\quad\text{and}\quad T_k\vdash s\}$$
$$T_i\sqcup T_k=\{s\mid T_i\vdash s\quad\text{or}\quad T_k\vdash s\}$$

(In the notation of Tarski (1956) the first set is $T_i\cdot T_k$, the second T_i+T_k.)

The former is the infimum (*glb*) and the latter the supremum (*lub*) of $\{T_i, T_k\}$. Hence we have proved

THEOREM 7.1 The structure $\mathscr{L} = \langle \mathscr{T}, \sqcap, \sqcup \rangle$, where $\mathscr{T}$ is the set of theories with a common core, is a lattice.

Let us now move on to matters of significance.

2.2. *Significance*

Let us begin by restricting our considerations to expressions belonging to a conceptual language $\mathscr{L}$ free from ambiguity. In this case the relation $\mathscr{D}$ of designation may be construed as a function from expressions Σ^{**} of $\mathscr{L}$ into constructs (see Ch. 1, Sec. 3.2). We stipulate that significance is a property that signs acquire when they happen to designate constructs – which is the case of numerals but not of musical notes. A sign of this kind signifies its significance, which is in turn the meaning of the construct it symbolizes. Briefly, signification (the function taking significances as values) is the composition of designation and meaning. In symbols: $\mathscr{S}ig = \mathscr{M} \circ \mathscr{D}$. More explicitly,

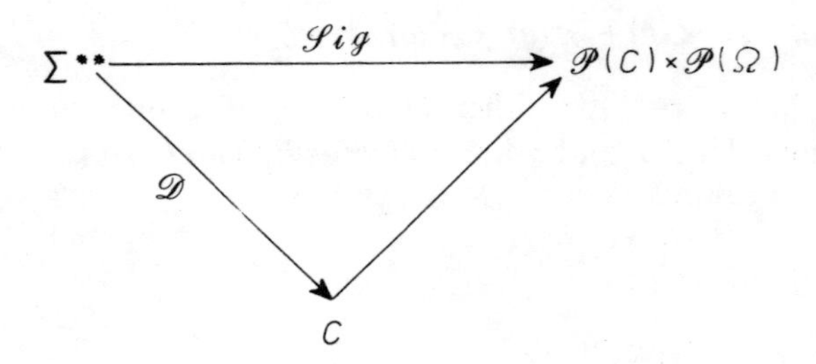

In other words, we are proposing

DEFINITION 7.4 Let Σ^{**} be the set of expressions of a conceptual language $\mathscr{L}$, $\mathscr{D}: \Sigma^{**} \to C$ the designation function, and $\mathscr{M}: C \to \mathscr{P}(C) \times \mathscr{P}(\Omega)$ the meaning function. Then the composition $\mathscr{M} \circ \mathscr{D} = \mathscr{S}ig$ is called the *signification function* of $\mathscr{L}$ and its value $\mathscr{S}ig\,(s, \mathscr{L})$ for any s in $\mathscr{L}$ is called the *significance* of s in $\mathscr{L}$.

COROLLARY 7.2 The significance of a sign belonging to a conceptual language equals the meaning of the construct designated by the sign, i.e.

If $\mathscr{D}sc$ in $\mathscr{L}$, and $\mathscr{M}(c)=\langle\mathscr{S}(c), \mathscr{R}(c)\rangle$, then $\mathscr{S}ig(s, \mathscr{L})=\langle\mathscr{S}(c), \mathscr{R}(c)\rangle$.
Proof By Definitions 1 and 4.

The first component or projection $\mathscr{S}(c)$ of $\mathscr{S}ig(s, \mathscr{L})$ may be called the *vicarious sense* of s, and the second coordinate $\mathscr{R}(c)$ its *vicarious reference*. These names convey the idea that, although signs are physical objects and therefore devoid of conceptual properties, if they proxy for constructs they indirectly acquire a meaning. This vicarious meaning is their significance. All this is expressed succinctly and exactly in the following diagram.

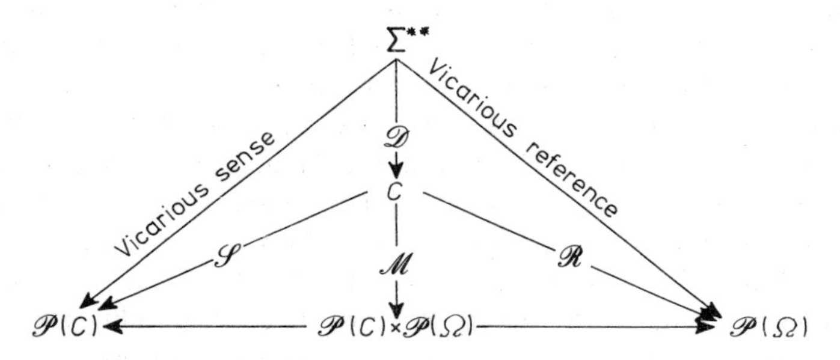

Table 7.2 illustrates these concepts.

TABLE 7.2

The dual significance of signs: four examples

Sign	Vicarious sense	Vicarious referents
$\circ$	Concatenation	Unspecified individuals
$+$	Addition	Numbers
V	Velocity	Physical systems
U	Utility	Objects and persons

All the previous considerations hold for a language in which each sign represents just one construct. (The converse need not hold: one and the same construct may be represented by two or more signs.) The language dependence of the signification function is made explicit in the following conventions.

DEFINITION 7.5 A sign s is *significant* in a language $\mathscr{L}$ iff s designates a construct in $\mathscr{L}$.

DEFINITION 7.6 A sign s is *nonsignificant* (or *syncategorematic*) in a language $\mathscr{L}$ iff s designates no construct in $\mathscr{L}$.

An isolated individual variable letter such as 'x' and a stray individual predicate letter such as 'F' are nonsignificant precisely because they are stray, i.e. outside any theory that may assign them definite senses. As Frege (1912) warned, it won't do to say that they have variable meanings or even indeterminate meanings. But if a sign belongs to the symbolism of a definite theory then it signifies even though its vicarious reference be indeterminate – as is the case with an abstract theory. For, in this case, the sense of the symbol is determined by the axioms of the theory in question – and sense is sufficient for meaning.

The last two definitions are relevant to the one-sign-one construct languages. These are unusual: most actual languages are ridden with ambiguities, i.e. they do not possess a designation function but rather a designation relation. For example, in informal mathematical contexts the definite integral sign may designate any of a dozen or so concepts of integral: Cauchy, Riemann, Stieltjes, Lebesgue, Schwartz, etc. The precise significance of any expression involving the integral sign will depend then on the precise interpretation assigned this ambiguous sign. (So much so that some such expressions make no sense whatsoever under certain interpretations of the snake symbol. For example, the Riemann integral of the Dirac delta is as nonsignificant as '1/0'.) In other words in most cases a language, even a mathematical one, will be the result of superposing two or more different languages equipped with one designation function each. When significance questions arise, the expert starts by analyzing the given mixture. And then he may use our previous definitions.

Note finally that a sign designating a construct with nil sense is significant. On the other hand a sign such as 'spinning and inconsistent' is insignificant because "spinning" and "inconsistent", are defined on disjoint domains. (Likewise a zero velocity is a velocity, whereas hopes have no velocity because the velocity function is not defined for them.) By contrast 'round square' is significant because both "round" and "square" are defined on the same domain, i.e. the set of plane figures. This is what makes it possible to disprove the statement that there are figures both round and square. Incidentally, extensionalism cannot cope with this fact.

2.3. *Significance Assignment*

Unlike meaning, significance is *assigned* to its bearers rather than being inherent in them. By itself, i.e. apart from a more or less precise significance assignment, a sign is just a physical object. Hence it is mistaken to ask 'What does x signify?'. One should ask instead 'What significance has x been assigned in the language $\mathscr{L}$?'. Or, in pragmatic terms, 'What are we supposed to think of or do at the sight of x?'. In other words, it is up to us whether and if so what a sign is to signify. Not so with the constructs symbolized: even though we need not conceive of them as Platonic Ideas, they must be regarded as being born with some meaning – for otherwise they are nothing. In other words, whereas symbols are conventional and therefore substitutable, the constructs they symbolize are subject to law – logical, mathematical, or scientific. Consequently the disclosure of significance rests on knowledge rather than being the subject of some "language game".

If we wish to find out the standard significance of an ordinary language expression we look it up in a dictionary. But if the expression belongs to a mathematical language or to a scientific symbolism, then one must look it up in the respective theory. In either case we come up with the vicarious sense and the vicarious reference of the sign. The two go hand in hand even if one or the other happens to be somewhat hazy. Thus when elucidating the significance of an operation symbol such as '+' we must bring in the set or sets on which the operation (not the symbol) is "defined" – and the members of this set are precisely the referents of the operation (a construct). Similarly with the elucidation of the significance of scientific terms: here too the determination of the (vicarious) referents of a symbol is part of the assignment of its (vicarious) sense. In particular, a predicate symbol will signify a property of the (vicarious) referent of the symbol. Therefore it might be thought that there is no point in keeping sense and reference as distinct components of meaning. It might be surmised that reference is a function of sense. But this is not quite so: (i) an axiom set such as Peano's, precise though it is, does not characterize completely its objects or referents but fits a number of them; (ii) a scientific theory initially intended to represent things of a certain kind may turn out to refer to a different kind of object. And even if sense were to determine reference in an unambiguous way, this would not invalidate our Definitions 1 and 4, for surely an ordered couple remains such if its second coordinate is determined by the first one – as is the case with the pair $\langle x, y \rangle$, where $y = f(x)$.

Keeping sense and reference distinct has definite advantages. In the case of mathematics it allows us to elucidate the differences in meaning among the various realizations of a given abstract formalism, namely thus. Here we have as many meanings as interpretations (or as models). The first coordinate of each such meaning consists of a fixed sense determined by the abstract theory enriched with the semantic assumptions that determine the particular interpretation. And the second coordinate of the meaning value is the domain of individuals, which varies from interpretation to interpretation. What is common to all these is of course the sense of the abstract theory. And in the case of factual science the distinction between the two meaning components is a useful reminder that one and the same thing may be represented by a number of constructs. Here, unlike the case of mathematics, the second coordinate is held fixed: the thing referred to does not change with the point of view. But in both cases we have to do with meaning differences. We shall come back to this problem in Sec. 3.3.

In mathematics and in science the assignment as well as the analysis of significance start with the rules of designation that match symbols with constructs. The second and most important step is to characterize the construct itself – and this is a matter of theory not of rule. The characterization ("definition") is often incomplete: sometimes because we want to leave room for ulterior specification, at other times because we do not know any better. The former is the case with abstract mathematics: a full specification of both sense and reference, i.e. a "definition" of the theory of a particular model, will destroy the leeway typical of abstract mathematics. In the case of science even if we wanted to effect a complete characterization we could not get it. Any characterization of the factual sense of a scientific construct is bound to be incomplete because it includes semantic assumptions that "point" to the referents without any further assistance from mathematics. When mentioning a specific domain of individuals, such as the real numbers, the mathematician can fall back on the specific theory or theories that "define" those individuals. Not so the scientist: he does not construct the referents of his theories but finds them or hopes to find them, hence he cannot fall back on other constructs.

The best he can do is to set up a mathematical formalism and conjoin it with semantic assumptions interpreting the basic concepts as so many factual items. The latter are at best described, at worst just mentioned. Whence the unavoidable meaning indeterminacy of factual theories.

2.4. *Degrees of Significance Definiteness*

There are degrees of significance definiteness corresponding to the degrees of meaning determinacy. We may distinguish the following degrees:

(i) *Low*

(*a*) Sense: definite but minimal. Reference: arbitrary except for the very general conditions set by the postulates determining the sense. Example: any abstract theory.

(*b*) Sense: not fully specified. Reference: definite. Example: any body of factual knowledge that does not contain full blown theories.

(ii) *Medium*

(*a*) Sense: minimal enriched with interpretations of some of the basic constructs. Reference: either definite or arbitrary, depending on whether or not the basic sets are specified. Example: any partial model in the sense of Ch. 6, Sec. 2.2, Definition 2.

(*b*) Sense: specified by a theory formulated in an intuitive or heuristic way. Reference: definite. Example: almost any theory in factual science.

(iii) *Large*

(*a*) Sense: nearly full. Reference: definite. Example: any theory in intuitive or nonformalized mathematics.

(*b*) Sense: specified by an axiomatic theory containing factual semantic assumptions. Reference: definite but large (genus not species of things). Example: any generic factual theory formulated axiomatically.

(iv) *Maximal*

(*a*) Sense: total. Reference: definite. Example: any fully axiomatized theory of a particular model.

(*b*) No instances in factual science.

We can now state the necessary and sufficient condition of significance. It is no more and no less than this: For a sign to be *significant* it must designate a construct (Def. 5). And for a construct to have a *reasonably definite meaning*, i.e. for being reasonably definite itself, the construct must belong to a reasonably well organized body of knowledge. Optimal meaning definiteness is attained only in a theory proper. (But maximal, i.e., full meaning is achieved only in mathematics.) While the profile of a construct can often be sketched in a satisfactory manner with more modest means, only incorporation into a theory takes automatic care of the syntax and, to some extent, of the semantics of a construct. For example mechanics will denounce 'The mass of that car is 3' as ill formed, for this expression fails to specify the mass unit. And it will expose 'Vorticity is noble' as a mongrel or semantically ill constructed expression, hence just as nonsignificant as the former expression. On the other hand ordinary language analysis has precious little to say about either except that they are grammatically sound. (Category mistakes are conceptual not linguistic: recall Ch. 2, Sec. 5.1.)

Finally a warning: We are not claiming that a construct is meaningless if it belongs to no theory, but only (i) that a construct has no *clear cut* meaning but in the midst of a theory, (ii) that a construct may *change* its meaning (i.e. it may become a different construct) if transplanted to a different theory, and (iii) that a *theoretical* construct exists only in a theory. All three aspects are illustrated in the following example. Even if different theories of social change start out with the same dictionary definition of "revolution", e.g., as "a drastic and sudden change in the established social patterns", they may elucidate or refine it in different ways, thus ending up with *different concepts* of revolution. They do this by assuming different actors (or referents of "revolution") and by emphasizing different traits as well as different causes. Thus while one theory will claim that the protagonists of revolutions (the referents of "revolution") are institutions, another will assert that they are social classes and a third that they are individuals. And while one theory will focus on institutional changes another will emphasize changes in the social and economic structure while a third view will concentrate on changes in individual roles. Finally, while one theory will assume that revolutions occur when institutions outlive their usefulness, another will claim that they are the last straw in a class struggle, and a third doctrine will assert that they happen when the members of the ruling class become corrupt. Clearly, 'revolution' signifies different constructs in the various theories of revolution even though they all share the core meaning assigned by the dictionary. (In many cases there is no such core meaning constituted by the pretheoretical or intuitive construct.) Tell me the company a construct keeps and I will tell you what it is: recall Ch. 5, Sec. 4.

3. Meaning Invariance and Change

3.1. *Synonymy*

Two signs will be said to be *semantically synonymous* just in case they have the same significance. And two symbols will be regarded as *pragmatically synonymous for a given user* if he employs them interchangeably or if, under the same circumstances, they evoke in him the same reactions. Pragmatic synonymy does not presuppose semantic synonymy: thus for many people 'psychiatrist' and 'psychoanalyst' are synonymous. And few of us are consistent in matters of pragmatic synonymy. If for no other reason, semantics cannot be based upon pragmatics. A second reason is that the determination of pragmatic synonymy calls for the observation of linguistic behavior, which is irrelevant to semantic synonymy: we do not circulate questionnaires in order to find out whether 'mass' and 'inertia' are semantically synonymous. We shall restrict our considerations to semantic synonymy.

Let us begin by restating our definition in more explicit terms:

DEFINITION 7.7 Two signs are said to be *synonymous* in a given language $\mathscr{L}$ iff they have the same significance in $\mathscr{L}$:

If s and s' are in $\mathscr{L}$, then $\mathscr{S}yn(s, s', \mathscr{L}) =_{df} \mathscr{S}ig(s, \mathscr{L}) = \mathscr{S}ig(s', \mathscr{L})$.

Example 'John loves Mary' and 'Mary is loved by John' are linguistically different but synonymous sentences: they are semantically identical – they express the same proposition.

COROLLARY 7.3 Synonyms designate the same constructs:

$$\mathscr{S}yn(s, s', \mathscr{L}) \text{ iff } \mathscr{D}`s = \mathscr{D}`s' \text{ in } \mathscr{L}.$$

COROLLARY 7.4 Synonyms have the same vicarious sense and the same vicarious reference:

$$\mathscr{D}sc \text{ and } \mathscr{D}s'c' \text{ in } \mathscr{L} \text{ and } \mathscr{S}yn(s, s', \mathscr{L}) \text{ iff } \mathscr{S}(c) = \mathscr{S}(c') \text{ and } \mathscr{R}(c) = \mathscr{R}(c').$$

Proofs The first corollary follows from Definitions 5 and 7. The second from Corollary 3 and the axiom for ordered pairs.

Remark 1 Corollary 4 looks pedantic but it is not, for we do not know that two signs do stand for the same construct, and are thus synonymous, unless we analyze their designata into sense and reference and prove, in a more or less rigorous way, that the senses and the referents are the same. A relatively frequent case is this: two different lines of reasoning within a theory come up with a function each. Further research shows that the two functions satisfy the same differential equation and are subject to the same initial or boundary conditions. This proves that the two functions are the same, or that they differ at most by a constant. *Remark 2* Logical equivalence is insufficient for synonymy. Not even equality may be enough: thus that two functions, f and g, share their values at a point a, i.e., $f(a)=g(a)$, does not entail that '$f(a)$' and '$g(a)$' are equisignificant or synonymous. Only identity guarantees synonymy. *Remark 3* If we demand that the definiendum and the definiens have the same meaning then we can accept only identities as qualified to define, since only an identity assures us that the two sides are just different names for one and the same object. Consequences: (*a*) equivalence is not the proper form of a definition, and (*b*) the intuitive asymmetry between definiendum (*LHS*) and definiens (*RHS*) is lost in the object language: it may be regarded as a metatheoretical or as a pragmatic trait. More in Ch. 10, Sec. 2.2.

As with synonyms so with antonyms:

DEFINITION 7.8 Two signs are *antonymous* in a given language $\mathscr{L}$ iff each designates the negate of the other's designatum:

If s and s' are in $\mathscr{L}$, and $\mathscr{D}sc$ and $\mathscr{D}s'c'$, then $\mathscr{Ant}(s, s', \mathscr{L}) =_{df} c' = \neg c$.

Remark By the double negation theorem of ordinary logic, $\mathscr{Ant}$ proves to be a symmetrical relation. If intuitionistic logic were adopted this would be lost unless the antonymy relation were redefined. Because we find no use for intuitionistic logic in science, we won't bother with this problem.

Synonymy and antonymy are just two bands of the entire spectrum of significance relations. Both $\mathscr{Syn}$ and $\mathscr{Ant}$ come in degrees: there is weak synonymy as well as weak antonymy. Thus 'set' and 'class' are weakly synonymous, and 'war' and 'truce' weakly antonymous. The following definitions elucidate the concept of likeness of significance.

DEFINITION 7.9 If s and s' are signs designating the constructs c and c' respectively, then s and s' are *partially synonymous* (or they exhibit a *likeness of significance*) iff the intersection of the senses of c and c' is not nil: $\mathscr{S}(c) \cap \mathscr{S}(c') \neq \emptyset$.

Example 'Hemisphere' in geography does not mean the same as in anatomy, but the senses of the two constructs concerned are close even though their referents (Earth and brain) are different.

DEFINITION 7.10 If s, s' and s'' are signs designating the constructs c', c' and c'' respectively, then the *likeness in significance* between s and s' is *closer* than the likeness in significance between s and s'' iff $\mathscr{S}(c) \cap \mathscr{S}(c') \supset \mathscr{S}(c) \cap \mathscr{S}(c'')$.

The preceding elucidations of the notions of equality and likeness of meaning should meet Quine's objection to propositions, namely that "If there were propositions, they would induce a certain relation of synonymy or equivalence between sentences themselves: those sentences would be equivalent that expressed the same propositions" (Quine, 1970b, p. 3, also 1960, Ch. VI). Well, don't they?

Our treatment of meaning relations, while at variance with the lexical view of semantics (Katz and Fodor, 1963), is in agreement with the view (Bar-Hillel, 1970) that such relations are logical relations. And our definition of synonymy confirms the view that the identity of the sets of implicants and implicates (i.e. sense) is necessary but insufficient for synonymy (Attfield and Durrant, 1973).

3.2. *Meaning Invariance*

The relation $\mathscr{S}yn$ of synonymy introduced by Def. 7 (Sec. 3.1) is an equivalence relation. Hence it defines the equivalence classes constituted by synonymous signs. In other words, for any language $\mathscr{L}$, if s is in $\mathscr{L}$,

$$[s]_{\mathscr{L}} = \{t \in \mathscr{L} \mid \mathscr{S}yn\,(s, t, \mathscr{L})\}$$

is the set of all the synonyms of s in $\mathscr{L}$. Every one of these equivalence classes corresponds to a single construct.

Take now the totality of equivalence classes under $\mathscr{S}yn$, i.e. the quotient set $\Sigma^{**}/\mathscr{S}yn$. This is the linguistic representative of all the constructs expressible in $\mathscr{L}$. Even if the designation function $\mathscr{D}$ is many-one, as we have assumed, we now have a one-one function $\mathscr{D}^*$ that matches distinct elements of $\Sigma^{**}/\mathscr{S}yn$ to distinct members of the set C of constructs expressible by $\mathscr{L}$. This isomorphism $\mathscr{D}^*: \Sigma^{**}/\mathscr{S}yn \to C$ may be called the *regular designation function*.

Let us pull the various items together. We start with the many-one function (surjection) $\mathscr{D}: \Sigma^{**} \to C$. Then we define the equivalence relation $\mathscr{S}yn$ in Σ^{**}. This relation determines the projection $p: \Sigma^{**} \to \Sigma^{**}/\mathscr{S}yn$ that assigns to every symbol the class of its semantic equivalents. Next we map the range of p onto C. Finally we compose p with $\mathscr{D}^*$. The outcome is $\mathscr{D}^* \circ p = \mathscr{D}$ as pictured by the diagram

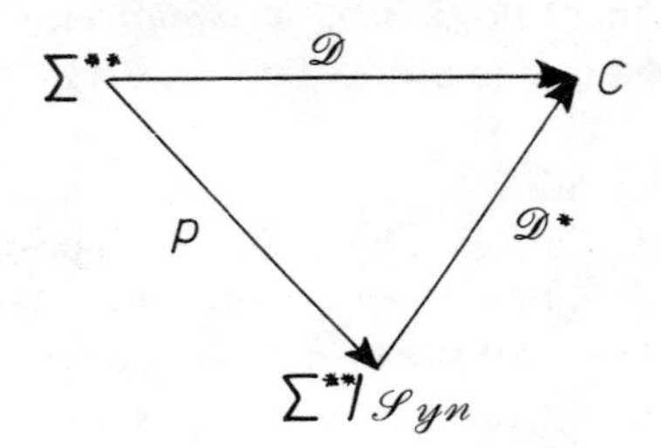

Now pick all the sentences of $\mathscr{L}$. Next group them into classes of synonyms, i.e., equisignificant sentences. Finally allow $\mathscr{L}$ to range over the set of all possible languages. That is, build the family of equivalence classes of sentences under the relation of equisignificance:

$\{[s]_{\mathscr{L}} \mid s \text{ is a sentence of } \mathscr{L} \ \& \ \mathscr{L} \text{ is a conceptual language}\}$. A nominalist might wish to identify this family of equisignificant sentences with what we call a *proposition* (or statement). But he cannot do this because he has no independent criterion of significance. In other words, the above cannot be treated as *defining* "proposition" because the latter concept is involved in the formation of the equivalence classes $[s]_{\mathscr{L}}$. Indeed, as we saw in Sec. 3.1, we do not know whether two sentences in mathematics or in science are synonymous unless we can show that they designate the same construct. Still, the above does clarify the idea that a proposition is that which remains invariant under all faithful sentence to sentence translations, as suggested by Russell (1940).

The translation concept can be elucidated in the following way. Consider the set $S_{\mathscr{L}}$ of all the sentences of a language $\mathscr{L}$ and the homologous collection $S_{\mathscr{L}'}$ for another language $\mathscr{L}'$. Even if these two sets are disjoint and structurally different, there may exist a one many relation

τ from $S_{\mathscr{L}}$ to $S_{\mathscr{L}'}$ that preserves significance. If such a relation exists we say that τ is a translation of $\mathscr{L}$ into $\mathscr{L}'$. More explicitly, we have

DEFINITION 7.11 Let $\mathscr{L}$ and $\mathscr{L}'$ be two conceptual languages and let τ be a one many relation from the set of sentences $S_{\mathscr{L}}$ to the set of sentences $S_{\mathscr{L}'}$. Then τ is said to be an *exact pointwise translation* of $\mathscr{L}$ into $\mathscr{L}'$ iff

$$\mathscr{S}ig(\tau`s)=\mathscr{S}ig(s) \quad \text{for every} \quad s\in S_{\mathscr{L}},$$

where $\tau`s\in S_{\mathscr{L}'}$ is a translation of s into $\mathscr{L}'$.

This concept of translation is useful in mathematics, where it can be strengthened to that of a function (Wang, 1951). But it is hardly applicable to natural languages, where not all sentences are independently meaningful. Here whole clusters of sentences must be paired off if significance is to be preserved. In other words, in the case of natural languages we must give up the ideal of pointwise translation and settle instead for global translation. However, this need not force us to adopt Quine's doctrine of the inevitable indeterminacy of translation: it only suggests supplementing Definition 11 with the following elucidation of the notion of global translation:

DEFINITION 7.12 Let $S_{\mathscr{L}}$ and $S_{\mathscr{L}'}$ be the sets of all sentences of languages $\mathscr{L}$ and $\mathscr{L}'$ respectively, and call $\mathscr{P}(S_{\mathscr{L}})$ and $\mathscr{P}(S_{\mathscr{L}'})$ their corresponding power sets. Then τ is an *exact global translation* of $\mathscr{L}$ into $\mathscr{L}'$ iff τ is a relation from $\mathscr{P}(S_{\mathscr{L}})$ to $\mathscr{P}(S_{\mathscr{L}'})$ such that

$$\mathscr{S}ig(\tau`u)=\mathscr{S}ig(u) \quad \text{for every} \quad u\in\mathscr{P}(S_{\mathscr{L}})$$

where $\tau`u\in\mathscr{P}(S_{\mathscr{L}'})$ is a translation of the set of u of sentences of $S_{\mathscr{L}}$ into $\mathscr{L}'$.

In sum, our concept of synonymy has allowed us to define two concepts of translation: pointwise and global. These concepts are strictly semantic as it should be: translation is concerned with meaning not with structure. (For the opposite view and a purely syntactic definition of translation see Svenonius (1973).) Surely perfect translations, even if global, are hard to come by. But, since they are desirable, the above definitions, far from being idle, may perform a regulative function. Thus it would be desirable to regularize the grammars and vocabularies of some of the natural languages so as to render them perfectly translatable into one another. Once this linguistic reform were implemented machine translation should offer no obstacles.

The concept of translation is relevant to linguistics, the foundations of mathematics, and the foundations of science – where it occurs with reference to equivalent theories utilizing different mathematical "languages". (For the use of theories as languages for further theories, see Ch. 1, Sec. 2.3.) However in factual science one is much more interested in different theories, whether or not they are couched in the same mathematical "languages". And, unlike translation, the passage from one theory to another may involve meaning changes. This matter deserves a separate subsection.

3.3. *Meaning Change*

If meaning is *sensus cum referens* then a change in meaning is a change in either sense or reference or both. And either change, when stripped of its pragmatic aspects, consists of a *difference* in sense and/or reference. Since both sense and reference have been constructed as sets, it is natural to define the difference in either as a symmetric (or Boolean) difference. It is equally natural to define difference in meaning as the pair ⟨difference in sense, difference in reference⟩. More precisely we have

DEFINITION 7.13 Let c and c' be constructs. Then
(i) The *difference in sense* between c and c' is
$\delta_{\mathscr{S}}(c, c') = \mathscr{S}(c) \Delta \mathscr{S}(c')$;
(ii) the *difference in reference* between c and c' is
$\delta_{\mathscr{R}}(c, c') = \mathscr{R}(c) \Delta \mathscr{R}(c')$;
(iii) the *difference in meaning* between c and c' is
$\delta_{\mathscr{M}}(c, c') = \langle \delta_{\mathscr{S}}(c, c'), \delta_{\mathscr{R}}(c, c') \rangle$.

The least interesting case is the one in which the meaning "distance" between two constructs is maximal:

COROLLARY 7.5 Let c and c' be semantically unrelated constructs, i.e. such that both $\mathscr{S}(c) \cap \mathscr{S}(c') = \emptyset$ and $\mathscr{R}(c) \cap \mathscr{R}(c') = \emptyset$, which amounts to $\mathscr{M}(c) \times \mathscr{M}(c') = \langle \emptyset, \emptyset \rangle = \square$. Then

$$\delta_{\mathscr{M}}(c, c') = \langle \mathscr{S}(c) \cup \mathscr{S}(c'), \mathscr{R}(c) \cup \mathscr{R}(c') \rangle.$$

The next case is far more interesting, as it concerns constructs with comparable senses:

COROLLARY 7.6 Let c and c' be two constructs such that the sense of c' contains the sense of c and such that they are coreferential. In brief, assume $\mathscr{S}(c')=\mathscr{S}(c)\cup\Delta\mathscr{S}$, with $\Delta\mathscr{S}\neq\emptyset$ and $\mathscr{S}(c)\cap\Delta\mathscr{S}=\emptyset$, and $\mathscr{R}(c)\cap\cap\mathscr{R}(c')\neq\emptyset$. Then

$$\delta_{\mathscr{M}}(c, c')=\langle\Delta\mathscr{S}, \Delta\mathscr{R}\rangle.$$

Example 1

$c=$ Partially ordered set $\langle A, \preccurlyeq\rangle$
$c'=$ Semilattice $\langle A', \preccurlyeq, \wedge\rangle$
$\mathscr{S}(c)=$ Axioms and theorems for posets, $\mathscr{R}(c)=A$
$\mathscr{S}(c')=\mathscr{S}(c)\cup$ Axioms and theorems containing $\wedge$, $\mathscr{R}(c')=A'\subseteq A$
$\delta_{\mathscr{M}}(c, c')=\langle$Axioms and theorems containing $\wedge$, $A-A'\rangle$.

Example 2

$c=$ Celestial mechanics (CM)
$c'=$ Lunar theory (L)
$\mathscr{S}(c)=$ Axioms and theorems of CM, $\mathscr{R}(c)=$ All celestial bodies
$\mathscr{S}(c')=\mathscr{S}(c)\cup$ Hypotheses concerning the Moon only, $\mathscr{R}(c')=\{\text{Moon}\}$
$\delta_{\mathscr{M}}(c, c')=\langle$Hypotheses concerning the Moon only, All celestial bodies except the Moon$\rangle$.

Whether or not our concept of meaning change elucidates the intuitive ideas put forth by Hanson, Kuhn, Toulmin and Feyerabend, and tossed so vehemently by so many philosophers, is hard to say. The preceding elucidations are offered as a semantic framework within which historical examples can be profitably discussed. Of course one may get inspiration for a theory of meaning by studying case histories: but such a study does not constitute an analysis of meaning, let alone a theory of meaning. Without a previous agreement on what 'meaning' signifies, i.e., unless a definite theory of meaning (and meaning change) is shared, if only for the sake of argument, the latter is bound to be chaotic, hence sterile. (For an example of such a dialogue among the deaf see the Minnesota discussion on correspondence rules, in Radner and Winokur (1971). For criteria for meaning changes see Kleiner (1971).)

The study of actual changes in significance belongs to pragmatics, historical linguistics, and the history of ideas. From this viewpoint every sign has a certain flexibility, also called *open texture* (Waismann, 1955). Thus "solid" has been redefined a number of times and will presumably be the object of further elucidations as the theories of solids evolve. It is only within the symbolism of a given theory that signs are nearly non-porous. Nevertheless the various significances assigned a scientific term in the course of its history often do have a solid core – namely the intersection of all its various significances. This core is not the "essence" of the sign: it may well consist of certain external characteristics. And it is often minute.

4. Factual and Empirical Meanings

4.1. *Definitions*

To elucidate the notion of factual meaning we blend the results of Ch. 2, Sec. 4.1 concerning factual reference with those of Ch. 5, Sec. 3.3 about factual sense. We thus get a specialization of Def. 1 in Sec. 2.1:

DEFINITION 7.14 Let c be a construct with a factual sense $\mathscr{S}_F(c)$ and a factual reference $\mathscr{R}_F(c)$. Then the *factual meaning* of c is defined as

$$\mathscr{M}_F(c) = \langle \mathscr{S}_F(c), \mathscr{R}_F(c) \rangle.$$

Example 1 $c =$ Electrodynamics, or e for short.

$$\mathscr{M}_F(e) = \langle \{\text{Law statements, meaning assumptions, etc., of } e\}, \text{Electromagnetic fields} \cup \text{Bodies} \rangle.$$

Example 2 $c =$ Mind concept, or m for short.

$$\mathscr{M}_F(m) = \langle \ulcorner \text{The internal activity of the brain} \urcorner, \text{Higher animals} \rangle.$$

Factual meaning should not be mistaken for empirical meaning. A construct may be said to have an empirical meaning just in case it refers at least partially to human experiences of some kind – e.g. perceiving, thinking, or doing. Thus while ⌜There are neutrinos⌝ is factually meaningful (and even true), it is empirically meaningless, for we have no experience of neutrinos. If a construct is empirically meaningful then it is factually meaningful – but not conversely. This principle of our semantics matches the metaphysical assumption that experience is a proper part of reality – namely the one involving experients. Both principles are of course at the center of a realist philosophy, of which more in Ch. 10, Sec. 3.3.

The difference between the factual and the empirical can be rendered slightly more precise by introducing the following conventions.

DEFINITION 7.15 A predicate $P: A \times B \times \ldots \times N \rightarrow S$, where S is a set of statements, is called *factual* iff at least one of the cartesian factors of the set on which P is defined represents a domain of factual items.

DEFINITION 7.16 A factual predicate $P: A \times B \times \ldots \times N \rightarrow S$ is called *empirical* iff at least one of the cartesian factors of the set on which P is defined is a set of sentient organisms.

DEFINITION 7.17 A predicate that is factual but not empirical is called *strictly factual* or *objective*.

Example Whereas "temperature" is strictly factual or objective, "hot" is empirical for having been defined on the set of ordered pairs thing-sentient being.

The rest is obvious. A statement is factual iff it contains at least one factual predicate, empirical iff it contains at least one empirical predicate, and strictly factual iff it contains factual but not empirical predicates. Similarly for sets of statements, in particular for theories.

Warning: Scientists sometimes call 'meaningless' what is actually meaningful but either uninteresting or false. For example, the runaway solutions of the equations of motion of classical electrodynamics are sometimes said to be 'physically meaningless'. As a matter of fact they are meaningful: they represent the motion of a self-accelerated point charge. Only, they are false. Moral: Dig out the concepts underneath the words.

4.2. *The Search for Factual Meaning*

When a scientific theory attains axiomatic maturity, the basic constructs determine the meanings of all others. This state of refinement is of course an outcome of a process of concept and theory formation that is anything but rule directed. In the preaxiomatic stages, i.e., in all theoretical domains save foundations research, the search for meaning is, like the search for hypotheses and theories, a guessing-trial-correction zigzagging. Even if the mathematical ideas are clear, their factual sense and occasionally even their referents are all too often imprecise at this stage. In sum, the semantics of a factual theory, i.e., its sense and reference, emerges gradually. It grows as a result of (i) solving more and more problems in the theory, (ii) enhancing the organization of the theory, (iii) relating the theory to other theories, and (iv) analyzing and evaluating the key constructs of the theory.

A typical situation in the search for the factual meaning of a theoretical construct is this:

(i) formulating a problem in the context of a given theory (very often an ill organized one);

(ii) detaching the mathematical component of the problem – i.e. formulating a problem in mathematics;

(iii) solving the mathematical problem;

(iv) wondering about the factual meaning of the solution.

This last task can be very hard, particularly in the preaxiomatic stages. Even if all the constructs occurring in the formulation of the problem have clear cut meanings, the solution may be quite unperspicuous: we may be able to "read" every symbol in it without however "making sense" out of the whole. For what we really are after is not a term by term interpretation: what we want to find out is what the solution *represents* – i.e. what aspects (e.g. properties) of the system it symbolizes and what facts (e.g. events), if any, it models. This is why J. C. Maxwell claimed that v^2, the square of a particle velocity, "has no distinct physical meaning", nor does mv^2, where m represents the particle mass (Maxwell, 1871). Actually the compound construct "mv^2" is perfectly meaningful in our large sense of 'meaning', since it is built out of individually meaningful constructs in a formally correct way. The trouble with it is not that it is meaningless but that it fails to *represent a definite property* of the system concerned. On the other hand v and mv do represent a property each and so does $\frac{1}{2}mv^2$. Why should the factor $\frac{1}{2}$ make such a huge semantic difference? Because it is $\frac{1}{2}mv^2$, not mv^2, that figures as an independent term (an addend) in a *law* statement of particle mechanics, i.e. in a theorem that is supposed to represent a natural pattern. In short, although both "mv^2" and "$\frac{1}{2}mv^2$" have definite senses and the same referent, the former is attributed no "distinct physical meaning" in the sense that it represents no particular property of the system concerned – and this in turn because it plays no role in any law statement.

The occurrence of a construct as an identifiable component in a law statement (e.g., as an addend) is then a good clue as to its factual meaning. No less and no more. Indeed, one and the same "quantity" (magnitude) can often be decomposed in different ways and these differences are only mathematically meaningful. Moreover, even if a construct occurs as a separate component in a law statement, we may fail to "read" it without further ado and so may have to resort to other procedures. For example, in a field theory a construct such as $\nabla \cdot V$, where V is a vector field, may not be "identified" or "recognized", i.e. assigned a "distinct physical meaning" right away. We may first have to integrate $\nabla \cdot V$ over a space region: by Gauss' theorem the outcome will be the flux of V across the boundary of that region, and this derived quantity may represent a property of the system. But even this may be insufficient: we may have to look for additional clues. A very fruitful one is dimensional analysis. Thus if the dimension of a magnitude X is LT^{-1} we may suspect that X represents the velocity of something. But then it may not.

Very often a theoretical construct with a perfectly definite referent has no manifest factual meaning – or, if preferred, its meaning is hidden. This is particularly the case with what may be called the *source magnitudes*. These are functions with definite reference classes and definite senses but representing no definite property of their referents although they generate a number of representing functions. Examples: (*a*) the various potential functions, whose gradients represent forces; (*b*) the partition function in statistical mechanics which, by mathematical manipulation, yields a number of representing functions; (*c*) the wave function and the statistical operator in the quantum theories. As long as a source function furnishes functions representing one property each we should tolerate, nay encourage them and defend them from the attacks of the operationist, who has no use for them because they are not directly measurable.

Gradually and somehow or other, both the mathematical formalism and its factual meaning mature to the point that the theory becomes ripe for axiomatization. In particular, the semantic assumptions can be stated explicitly and so contribute, alongside the remaining assumptions, to delineating the factual meaning of the theory. Once the theory has been given an axiomatic formulation everything should run smoother than before. For in a well organized theory everything flows from the axiom top: both theorems and meanings. In principle the meaning of a defined construct will be found by analyzing its definition and the meaning of a theorem will be found by analyzing its component constructs as well as the premises that entail it. In principle but not necessarily in practice: axiomatization eases proofs and interpretations and makes them more precise but not mechanical.

4.3. *Shape and Role of Meaning Assumptions*

There is no consensus on how to handle the factual meaning of scientific constructs. Every scientist proceeds in his own way, though often influenced by some philosophical school. Here follow in quick succession the main current views on the matter – or rather sketches of them.

(i) *Formalism and conventionalism.* No meaning assumptions are called for because scientific constructs have no factual meaning: they are just pieces in a mathematical machinery. A scientific theory is identical to its mathematical formalism. *Criticism* Any mathematical formalism may be assigned alternative interpretations: a factual construct is a mathematical construct together with a factual interpretation (Ch. 6, Sec. 3).

(ii) *The semantic miracle view.* Meanings take care of themselves: every formalism generates its own interpretation. Hence no meaning assumptions are needed. *Criticism* Same as for (i). The reason some formalisms seem to be necessarily associated with certain interpretations is just habit. A scientific specialist is so used to handling certain form-content pairs that it may not occur to him that the same form may be paired to an altogether different content.

(iii) *The meaning-lies-in-the-name view.* All we need to turn a mathematical formalism into a scientific theory is to add designation rules, such as ⌜The parameter t is called time⌝. *Criticism* While designation rules are necessary ingredients of the semantics of a scientific theory, they are insufficient. Names are conventional, meaning assumptions are not: the latter are testable, hence capable of being overthrown. Thus although we still use the electric current concept we no longer assume that it represents the rate of flow of the electric fluid. If we claim that a theory represents certain things then we must state clearly what in the theory represents what in the world.

(iv) *Classical empiricism.* Meanings are assigned by ostensive definitions or rules, such as "That is blue". *Criticism* First, an ostensive rule may be a subject for pragmatics but not for semantics: unless accompanied by suitable gestures it is devoid of (pragmatic) meaning. Second, unfortunately for toddlers, the most interesting constructs in science and even in ordinary knowledge are non-ostensive.

(v) *Operationism.* Meanings are assigned by specifying modes of observation, measurement, or action in general. For example, the concept of a thermodynamic state must be specified by describing a method of preparation of states (Carathéodory, 1924; Giles, 1964). *Criticism* First, most of the referents of a scientific theory are beyond the reach of the experimentalist – if only because they are possibles rather than actuals. Hence if a thermodynamic state must be the outcome of a human act then most physical systems are in no state at all. Second, this view is dangerously near subjectivism, which has no need for empirical procedures. Indeed, once a thermodynamic state has been made to depend on man, why not on mind alone? This step has actually been taken: it has been claimed that "a state is a state of mind induced by the available knowledge of preparation" (Burton, 1968). Third, the whole of operationism rests on a confusion between reference and evidence (Feigl, 1958; Bunge, 1967a, 1973a, 1973b).

(vi) *Mellowed operationism.* A theoretical term is significant to the extent to which it is related, in the body of a theory, to some observational sentence of the theory (Carnap, 1956). More precisely, "A theoretical term t is significant if there is an assumption A involving t such that from A and additional assumptions involving other theoretical terms that have already been recognized as significant it is possible to derive with the help of the postulates and the rules of correspondence an observation sentence that cannot be derived without the assumption A" (Carnap, 1963a, p. 80). *Criticism* First, most scientific theories contain no observational constructs. Strictly speaking *all* the constructs in a scientific theory are theoretical and often devoid of empirical (yet not of factual) meaning. Hence the theoretical/observational dichotomy is not applicable to scientific theories. Second, as suggested by Carnap himself, testable statements can only be had with external help – namely by roping in assumptions belonging to other theories as well as empirical evidence. (For details see Bunge (1967a) and (1970a).) Even so, this will render the theory empirically testable but not empirically meaningful.

(vii) *Realism.* Factual theories have a factual meaning. And factual meanings are determined jointly by all the assumptions of the theory – in particular, though not exclusively, by the semantic assumptions. Since the latter bear on the basic concepts, they belong in the axiomatic foundations of the theory. Every semantic assumption indicates the referent(s) of the construct and suggests what it represents. For example, a conditional probability value $P_n(b \mid a)$ can be made to represent the tendency of the nth system of a certain kind to jump from state a to state b. It makes no difference to the semantics of a theory whether the referents are perceptible and the traits that are being represented by the theory are directly observable or must be reached with the help of further theories: observability is relevant to testability not to meaning. In other words, the semantic assumptions in a scientific theory relate constructs to things and some of their traits: anything that relates constructs to empirical operations, such as preparation or measurement, may qualify as a *testability condition* but certainly not as a meaning assumption. *Criticism* None that I can think of.

Every one of these policies has its advocates and all of them but classical empiricism (ostensivism) are practised nowadays by scientists. Yet popularity is not the seal of truth: just because most scientists either do not care to formulate any explicit meaning assumptions or because they propose operationist interpretations, does not prove that those policies are right. A semantic policy, just as any other policy, must be judged by its success in achieving the goals it sets itself as well as by its success in surviving criticism. If judged in this dual way, the most popular semantic policies prove to be utter failures. The first three because they do not even attempt to pinpoint the peculiarities of factual theories *vis-à-vis* their mathematical formalisms. And the two varieties of operationism are flops as well because the semantic assumptions they recommend are too narrow: they are linked to particular laboratory techniques, hence closed to possible alternatives. And most of the time they are phony, as they stipulate impossible measurements. Try measuring a lagrangian density, or a partition function, or a state function.

The failure of the various attempts to specify the meanings of theoretical concepts by reducing, or at least linking them to observational concepts, has produced a mood of despondency among philosophers of science. Thus Putnam came to the conclusion that the very *problem* of interpreting theoretical terms "does not exist" (Putnam, 1962) and Hempel thinks now that the problem was "misconceived" (Hempel, 1970). Nevertheless the problem won't lie down: the theoretical scientist faces it every day when pondering over the possible factual meanings of his mathematical formulas. Moreover the scientist tries to solve his semantic problems in more of less naive ways, without the benefit of a full fledged semantic theory. No doubt he would fare better if, instead of being told that he should worry no more because he has really no problem at all, he were offered a definite semantic theory.

The failure of empiricist semantics does not entail the impossibility of every kind of semantics. It just suggests looking elsewhere for a solution to the genuine and difficult problem of specifying (or rather sketching) the meaning of theoretical concepts (or the significance of theoretical terms). An alternative to empiricism is realism, the only semantic policy that has emerged unscathed from the fifty years' war over the soul of scientific constructs. True, realism is not often practised in an explicit manner, i.e., by laying down the rules of denotation and the semantic assumptions that delineate the meanings of the undefined concepts of a scientific theory. But then (*a*) impopularity is not the seal of falsity, (*b*) few theoreticians care to render all their assumptions explicit, and (*c*) positivist semantics, even after having been repudiated by its propounders, still enjoys a considerable prestige among scientists. It behooves the philosopher to illumine the path by showing in particular cases how to interpret theoretical concepts in terms of facts.

We close with a warning. A semantic assumption occurring in an axiom system could rightly be called a *meaning postulate*. Unfortunately this expression was preempted by Carnap (1952), who employed it in a different sense. Consider the two standard examples of a "meaning postulate", first of all

⌜For every x, x is a bachelor iff x is a male and x is not married.⌝ (1)

This is a construct-construct relation that does not intend to say what "bachelor" refers to and represents but rather how it is related in extension to both "male" and "married". It seems to be an ordinary dictionary definition. In any case it is supposed to be a priori, hence empirically unassailable. The second standard example of a "meaning postulate" is

⌜For every x, if x is a bachelor then x is not married.⌝ (2)

This might be taken as a "law" of ordinary knowledge or as a linguistic fact or, finally, as a deductive consequence of the convention (1). In neither case is it an empirically refutable semantic assumption or axiom in our sense. Nor is it a postulate in the sense of 'axiom' (Carnap, 1952). Whatever it is, a Carnapian "meaning postulate" does not contribute to the sense of a scientific term, hence it takes no part in the axiomatic reconstruction of a scientific theory, whence it need not concern us.

5. Meaning et alia

5.1. *Meaning and Testability*

We have elaborated the common sense or realist view of factual meaningfulness. According to this view a statement is factually meaningful only if it refers to, and the more so if it represents, a factual item – a thing, a state of affairs, or an event. The referent need not be actual and the representation, if there is any, need not be true: the statement may well concern the past or the future, and it may be utterly false or even impossible to test. Any narrower criterion of factual meaningfulness runs the risk of disowning the most interesting scientific speculations.

This commonsensical view was also the one Carnap held before he came under the influence of Wittgenstein: "The meaning of a statement lies in the fact that it expresses a (conceivable, not necessarily existing) state of affairs" (Carnap (1928) in Carnap, 1967, p. 325). Eventually

Carnap and the other members of the Vienna Circle succumbed to Wittgenstein's verifiability doctrine of meaning, according to which the meaning of a sentence consists in the way it can be verified – i.e. in its truth conditions. (Cf. Schlick, 1932/33.) If a sentence belongs to empirical science, the truth conditions must describe procedures of empirical test. Whence *Meaning* = *Testability*. And because only observational statements are empirically testable, *Meaning* = *Observability*. Put negatively: Whatever sentence is not verifiable by observation is empirically (= cognitively) meaningless.

This strict thesis was somewhat liberalized later on but without relinquishing the dependence of meaning on testability. The criterion that ultimately prevailed in the logical empiricist camp was this: A sentence is empirically (or cognitively) significant only if its sole extralogical constants are observational (e.g. 'sticky' and 'smelly') or if, conjoined with further sentences, it entails observational sentences (Carnap, 1956, 1963a; Rozeboom, 1962). In the latest version of this thesis testability need not be scientific hence objective: "I regard as meaningful for me whatever I can, in principle, confirm subjectively" (Carnap, 1963b, p. 882). Once again the subjectivist component of empiricism won out.

We criticized earlier the identification of meaning with testability (Ch. 4, Sec. 3.3). Suffice here to remark that false statements are as meaningful as true ones, and that all scientific theories contain incompletely testable or even wholly untestable statements: for example, quantum mechanics allows one to compute the velocity of an electron in an atom, which is an empirically inaccessible function. Let us undertake to state, even if in summary, the actual relations between meaning and testability.

It is a rule of scientific method to abstain from assigning truth values, except for the sake of argument, until after the relevant empirical evidence is at hand. In other words, testing is necessary for truth value: *Truth value* ⇒ *Testing*. In turn, a necessary condition for any actual tests is that the statement concerned be testable, i.e. that the theories and the empirical techniques of the day judge the statement to be susceptible of being confronted with facts, if not right now later on. In sum, *Testing* ⇒ *Testability*. Now, if a statement is susceptible to test then it is meaningful to begin with, i.e. it has a nonempty sense and a reference class – which may, alas, prove to be empty. Otherwise it would be impossible to devise a test for the statement. Briefly, meaningfulness, though insufficient for testability, is necessary for it: *Testability* ⇒ *Meaningfulness*. Finally if a statement is factually meaningful, i.e., if it has a factual sense and a factual reference (whether actual or possible), then it is well formed, i.e. syntactically meaningful in some formalism. (For a definition of the latter notion of meaningfulness see Tarski (1956), p. 284.) In short, *Meaningfulness* ⇒ *Well-formedness*.

In conclusion, the complete *logical* chain is

> *Truth value* ⇒ *Testing* ⇒ *Testability* ⇒ *Meaningfulness* ⇒ *Well-formedness*.

Consequently the *methodological* sequence is

> *Checking well-formedness* → *Assigning or analyzing meaning* → → *Judging testability and devising tests* → *Actual testing* → *Assigning truth value*.

5.2. *Meaning and Use*

At about the same time as Carnap was working out the *Tractatus*' view on meaning, Wittgenstein was busy demolishing it and attempting to sketch a pragmatist philosophy of language. On this view language is only a social activity and the meaning of an expression consists in its use. In turn, uses are established by custom as recorded by the (Oxford) dictionary – not by any theoretical analysis. As one of the disciples put it, "To give the meaning of an expression (in the sense in which I am using the word) is to give *general directions* for its use in making true or false assertions" (Strawson, 1950). Therefore "Ultimately a meaning-statement (a statement as to what a linguistic expression means) is to be tested by determining what people do in their employment of the expression in question" (Alston, 1968, p. 145).

What the second Wittgenstein and his apostles were interested in was, of course, the notion of *pragmatic meaning*. This concept can be exemplified but has so far escaped theoretical elucidation. (Only the notion of pragmatic synonymy has been elucidated – not by the Wittgensteinians but by Carnap (1939) and Naess (1956).) In any case it is different from the concept of semantic meaning and no possible substitute for the latter. The same "general directions" for manipulating a bunch of symbols are consistent with alternative significations assigned to those symbols. For this reason linguistic pragmatism cannot explain why two quantum physicists, while disagreeing on the significance of the symbols they use, can arrive at the same formulas. While linguistic pragmatism is incompetent to tackle semantic meaning, it can be effective in misleading philosophers into thinking that meanings can only be discovered by listening to (ordinary) talk rather than by uncovering sense and reference. Surely field observations can disclose pragmatic meanings – provided they are conducted with the methodological apparatus of the linguist. It is linguists not philosophers who are equipped to carry out linguistic investigations. Philosophers should philosophize about linguistics among other things – not language.

5.3. *Meaning and Understanding*

The significance of a sign and the meaning of the construct it designates must not be mistaken for the mental process of understanding either. Meanings are supposed to be objective whereas the experience of thinking of them is, like any other experience, subjective. This distinction, which goes against the grain of empiricism, goes back to Bolzano, Lotze, Frege, and Meinong. It is consecrated by the distinction, emphasized in the Introduction to Volume 1, between philosophical semantics and the psychology of cognition and language, which is a branch of factual science.

The distinction between meaning and the understanding of meaning is correct provided meanings are not reified or turned into Platonic Ideas. With meanings as with their carriers, i.e., constructs, we may adopt a fictionalist stand: we may *pretend* that they exist – without however assuming that they have an *autonomous* existence. No rational beings, no constructs; no constructs, no meanings. We have no use for the idealist thesis that there are concepts and propositions in themselves, i.e. apart from thinking beings. But if we wish to build a *theory* of constructs that brushes aside the particular ways and circumstances in which constructs are being thought, then obviously we must abstract from such ways and circumstances. Such a theory cannot possibly enter into conflict with any psychological theory concerning the understanding (or the misunderstanding) of a meaning, since it will not pose the question.

Although the concepts of meaning and of understanding belong to different fields of research, they are related, namely thus: *If anything is understandable then maybe it is meaningful.* ('Maybe' because any claim to intelligibility is of uncertain value.) Practical consequence: The clearer and more orderly the presentation of a body of knowledge the better the chances of understanding it. It does not follow that axiomatics is the ideal didactical wrapping. It does follow that a judicious combination of axiomatics with intuitive remarks and a statement of motivations is the best one can do to *facilitate correct understanding*. (See Bunge, 1973b, Ch. 8, Sec. 6.)

Not surprisingly, recent work in psycholinguistics and in artificial intelligence bears out our thesis that isolated sentences are nonsignificant, hence unintelligible. In fact to understand a sentence a person (or a computer) must know the language the sentence belongs to, must be able to do some reasoning, and must have some substantive information (Winograd, 1972).

A last question: Can computers grasp any meanings? The short answer is: No, because they handle physical signals not constructs. It is the programmer who assigns to these signals definite constructs, hence the meanings of the latter: he does this when setting up or using his programming code. In particular when he reads (interprets) the computer output – or, for that matter, when he reads anything. Unlike its programmer, the computer does not have to and cannot possibly interpret anything. So much so that a computer is incapable of making interpretation errors: only rational beings can make semantic mistakes. Computers are used not because they replace minds but because they simulate some aspects of the human mind. Only a living brain can have a whole mind of its own. And only some brains pose new conceptual problems, invent theories, and evaluate the latter. For a computer to understand a sign (e.g. a sentence) consists in associating it to the correct state (of the machine) to produce a final output (or machine state) according to a definite rule included in the program. No understanding of meaning is involved in this purely physical operation.

5.4. *Factual Meaning and Covariance*

Place words like 'here' and time words like 'now' do not have the same significance for everyone: they are subject-dependent or egocentric. Similarly space and time coordinate values are local not universal. Hence a statement such as ⌜Particle p is at place x at time t⌝, even if true relative to some reference frame, is not universally true, i.e., true in, or relative to, every possible reference frame. (But if we know how the given frame is related to some other frame then we can translate the statement into another statement that will hold in, or relative to, the new reference frame. The Galilei and the Lorentz transformation formulas are two such translation devices. Recall Ch. 3, Sec. 2.3).

In other words, position and time values are not *invariant* under every coordinate transformation. (On the other hand electric charge and transition probability values are invariant.) And statements involving position and time coordinates are not always *covariant* (or form invariant) under certain coordinate transformations. In either case, what is at stake is the permanence, or lack of it, under changes in the mode of representing or mapping facts in spacetime: invariance (or noninvariance) in the case of properties, covariance (or absence of it) in the case of law statements. For example, a Galilei covariant formula holds in every Galilei frame, i.e. it is covariant under the group of Galilei transformations. This is the case with Newton's basic equations of motion, not however with their solutions: the latter are frame-dependent – but at least we know how to translate them into statements that are true in, or relative to, alternative frames. But Newton's equations are not covariant under different coordinate transformations such as Lorentz'.

It is tempting to conclude that only invariant properties and covariant equations have a factual or objective meaning, all others being endowed with a subjective meaning if any. Thus Weyl (1919, p. 129) held that a relation between points in spacetime has an objective meaning (*objektive Bedeutung*) just in case it is invariant under Galilei's transformations. But then Galilei's law of falling bodies, a special solution of Newton's equations of motion, would be devoid of objective meaning because it is frame-dependent (not covariant). This is an abuse of the word 'meaning'. What is really at stake is *truth* not meaning: A statement involving spatial or temporal concepts (e.g., spacetime coordinates) and making explicit reference to some frame can be fully meaningful without being *universally true*, i.e., true in (relative to) every possible frame.

But what is invariant (or else covariant) under a certain transformation group may cease to be so under a different group. Thus while the relation of simultaneity is Galilei invariant, hence absolute in Newtonian physics, it is assumed to be frame-dependent, hence relative, in relativistic physics. In the latter context, therefore, a statement of the form ⌜Events *a* and *b* are simultaneous⌝ is regarded as being ill formed unless the context makes it clear which reference frame has been adopted. Relative to a certain frame f it may be that the two events are simultaneous, but then they may fail to be simultaneous relative to an alternative frame f'. Consequently the corresponding statements will not be universally true but instead locally so. In obvious symbols: while ⌜$S(a, b, f)$⌝ may be true, hence meaningful, ⌜$S(a, b, f')$⌝ may be false, hence meaningful as well. On the other hand Weyl inferred that simultaneity has no objective meaning (Weyl, *op. cit.*, p. 146). Which, in addition to being a semantic mistake, contradicts his earlier statement that Galilei invariants are meaningful.

Hilbert went even further by demanding the covariance of all statements, basic and derived, theoretical and experimental, under the most general coordinate transformations, i.e., those occurring in general relativity. In fact he stated that "a proposition that is not invariant under every arbitrary transformation of the coordinate system must be regarded as *physically meaningless*" (Hilbert, 1924, p. 274). Equivalently: "a proposition is invariant and has therefore a physical meaning if it holds for every arbitrary coordinate system" (Hilbert, *op. cit.*, p. 278). As with the cases of Galilei and Lorentz invariance discussed a while ago, what is actually involved is a criterion of universal (or frame-independent, hence observer-free) *truth* not one of factual *meaning*.

Hilbert, and Weyl before him, may have been using the term 'meaning' in a colloquial sense for, taken literally, their sentences concerning objective meaning are meaningless. In any case they should not be construed as formal definitions of "meaning" as covariance. (See however Suppes, 1967.) If they were so construed then almost every physical statement would have to be ruled out as meaningless. Only the *basic* physical law statements, such as the variational principles and their immediate consequences, can be required to be covariant with respect to certain transformation groups: their solutions *must* be frame dependent (relative) if they are ever to be subjected to experimental tests, for test devices are in the habit of being attached to some frame or other and of yielding results that are seldom the same for every other frame (Bunge, 1961a, 1967b).

Examples like the following have been offered in support of the claim that covariance (of some kind) is required for objective meaningfulness. ⌜The car is at rest⌝ is not an invariant (or frame independent) statement, for the same car is really moving relative to almost every reference frame other than our planet. That statement is an instance of an incomplete or ill formed statement. The complete, well formed statement is ⌜The car is at rest in (or relative to) the ground⌝ (or the ferry or whatever may be the case). The latter statement is unobjectionable: it is meaningful and perhaps even true. But it is not generally covariant: it holds (is true) relative to one frame only. To disqualify ⌜The car is at rest⌝ we do not need a requirement as severe as general or even Lorentz or even Galilei covariance. Our criterion of meaningfulness in Sec. 2.3 will suffice, for the predicate "is at rest" does not belong to any body of contemporary factual knowledge: the concept of velocity, in particular zero velocity, involves the one of reference frame. Indeed, the velocity function is a certain function on the set of ordered triples: physical system p – reference frame f – velocity unit u. Hence while ⌜$V(p, f, u)=0$⌝ is well formed and meaningful, ⌜$V(p)=0$⌝ is not. This simple solution to the problem squares with actual scientific practice. And it is far more economical than replacing ordinary logic by some system of three valued logic (true, false, and meaningless) in order to accommodate freaks like ⌜$V(p)=0$⌝ or ⌜The mass of c equals 5⌝, as has actually been suggested (Suppes, 1959, 1965, 1967). As emphasized in Sec. 2.3, scientific constructs are not to be judged, let alone tampered with, in isolation from the theories they belong to, for the simple reason that only such theories exhibit their form and content. In particular scientific theory, not philosophy, is competent to determine whether or not (*a*) a given formula in scientific discourse is well formed and meaningful and (*b*) a given magnitude (or physical quantity) is frame independent or unit free or both (Bunge, 1971a).

In conclusion, invariance and covariance have nothing to do with meaningfulness, nor even with objectivity. Most of the known, meaningful, and even partially true formulas of physics are not generally covariant. Covariance, then, is not necessary for meaningfulness. Nor is it sufficient, as it is possible to propose any number of formulas that are covariant with respect to certain transformations but are devoid of factual meaning. Consequently covariance does not define meaningfulness. On the other hand covariance is necessary (though insufficient) for frame independent (hence observer free) truth, a condition only a few basic law statements satisfy. (See Bunge (1959b) and (1967b).)

6. Concluding remarks

Our view on meaning and significance combines sense with reference. Thus a term such as 'electron' signifies both "the lightest material unit of electric charge" and any concrete instance of this predicate. It therefore consecrates and systematizes the ambiguity of the word 'meaning' in ordinary languages – without however conflating the two meaning components.

In our semantics there is no separate theory of meaning. But for the meaning calculus sketched in Sec. 2.1, our theses on meaning are mostly definitions and they make sense only against the background of the theories of sense and reference expounded in Part 1. Once the latter are assumed our definitions of meaning, meaning change, significance, synonymy, and their kin, look natural and even trivial. In point of fact our definition of meaning as sense together with reference cannot compete in boldness with any of its rivals: that meaning is verifiability, or truth condition, or use, or understanding, or information, or general covariance, or what not. We are proposing the cowardly view that meaning is just meaning.

So far we have been concerned with that which makes a construct what it is – namely its sense and its reference. Change either and a different construct will emerge. Not so truth and extension: these bear on fully meaningful constructs of certain kinds and, if factual, they are determined *ab extrinseco* rather than by analysis. In particular, whereas every factual statement is born with a sense and a reference, it is assigned no truth value until after having been tested for truth (Sec. 5.1). Moreover every truth value assignment is corrigible in this case: the same factual statement may now be assigned a truth value, now another. This is why we could not tackle the problem of truth before. But now the moment of truth has come.

CHAPTER 8

TRUTH

Logicians and mathematicians may be envied (or perhaps pitied) for needing a single concept of truth, namely that of formal truth. And they must be commended for having subjected it to a rigorous theory, namely Tarski's, now incorporated into model theory. Moreover, since in this theory the concept of (formal) truth is derivative (definable in terms of satisfaction in a model), formal scientists need not regard it as basic. Besides, this concept of truth raises no questions of confrontation with any factual facts: in logic and mathematics checking and proving are purely conceptual operations.

Factual scientists and the semanticists of factual science do not have it so easy. They have to contend with a radically different concept of truth, as suggested by the typical statement ⌜Theory T is an approximately true representation of factual domain F (or a good approximation in view of the evidence concerning F)⌝. What is at stake here is the concept of *partial truth of fact*, which – in contrast to that of complete formal truth – does not occur in the semantics of formal science. In matters of fact there is not only factual reference but also, and consequently, adequacy or inadequacy with respect to fact. Moreover such adequacy or inadequacy is rarely complete: unlike logical truth and logical falsity, factual truth and factual falsity are not polar opposites but contraries since, though incompatible, they are not exhaustive. This is sometimes not realized because science employs ordinary logic, hence seems to espouse a two valued theory of truth. But logic is the theory of deduction not the theory of truth. It is perfectly possible to process logically a set of statements that are neither fully true nor completely false: this is what science does. Hence we need a theory of partial truth of fact consistent with ordinary logic. In the present chapter we shall explore a chromatic (many valued) theory of truth matching with black-and-white logic.

1. Kinds of Truth

1.1. *Truth Bearers*

Granting that "truth" makes sense at least in some contexts, the very first questions we must try to answer are: (i) what kind of objects are the bearers of truth and (ii) what kind of object is truth itself. There is of course a number of views on this matter. All of them agree that sentences are involved in these questions and that sentences are physical objects (strings of sounds or written marks) belonging to some language or other. But the various views assign sentences different roles in relation to truth. Table 8.1 exhibits the main traits of the principal views. (We have not included the "no truth" view of truth, set forth by Ramsey (1931), for being so glaringly at odds with the practice of testing for truth.)

The first view, naive idealism, is entrenched in the Western modes of speech and thinking. It seems right in holding truth to be a property of propositions (as distinct from judgments and sentences), wrong in regarding propositions as eternal objects which man can only "discover". But at least this phantasy of the autonomously existing propositions is sometimes heuristically fertile. Thus a mathematician may claim that a theorem that has not yet been formulated, let alone proved, is true – and he may set out to "discovering it". The obvious criticism is that nonexisting objects have no properties. Just as my unborn great-grandson is neither here nor there, so unthought formulas are in no theory (except potentially) and a fortiori they are neither true nor false. Moreover, existing statements that have not yet been proved are conjectures that should be assigned no truth values except for the sake of argument. Likewise a factual hypothesis cannot be said to have been true (or false) from all eternity, even before it was formulated. We can assign it a truth value only after having subjected it to tests – and even so the assignment may be provisional. But the proposition must have been stated to begin with.

In conclusion we reject naive idealism. We retain however the idea that constructs, e.g. propositions, are distinct from both thoughts and sentences, and that they can be true or false. But instead of assuming that there are propositions in themselves we assume that, to exist, propositions must be thought or uttered (or written) by some rational being in some language or other. And rather than claiming that truth and

TABLE 8.1

Main views on the nature of truth and its bearers

View	Sentences	Propositions	Truth
Naive idealism, e.g., Platonic "realism".	Express some propositions (the known ones).	Timeless objects existing by themselves	Intrinsic property of propositions. These are *T* or *F* whether we know it or not.
Neoidealism, e.g. Frege and the early Russell.	Express all propositions.	The meanings of sentences.	The object denoted by a true sentence.
Naive materialism, e.g. nominalism.	Physical objects with a factual reference and a truth value.	There are none.	Property of sentences in relation to their denotata.
Conceptualist materialism.	Express all propositions. Devoid of semantic properties except vicariously, i.e., via propositions.	Indispensable fictions – equivalence classes of certain thoughts.	Assignable to some propositions on the strength of others. Factual truth: relative, partial and ephemerous.

falsity are inborn, we assume that they can be attributed (in some degree) – as well as withdrawn. But we shall not elaborate on these psychological presuppositions: we shall take them for granted.

The second view, neoidealism, shares with Platonism the belief that there are constructs in themselves. But, unlike Platonism, it is muddled and even inconsistent. First truth and falsity are predicated of sentences (or of propositions, according to the way the ambiguous German *Satz* is translated). Next they are made into separate Platonic entities, *das Wahre* and *das Falsche.* Worse, truth value is equated with the referent, or nominatum, or designatum (*Bedeutung*) of the sentence (Frege, 1892). Thus the *Bedeutung* of ⌜Chimpanzees are clever⌝ will be said to be the same as that of ⌜$2^2=4$⌝, namely The Truth. This is not only a confusion but it renders the concept of reference unnecessary, as it makes all true sentences have the same "referent" (*Bedeutung*) regardless of their genuine referents. Frege could obviously afford to make a couple of mistakes. (Another mistake of his was to oppose the type of axiomatics initiated by Peano and Hilbert as well as the correlative notions of definition by postulates and conditional definition.) Unfortunately Frege's muddle has been taken seriously by some of the best logicians and philosophers of our time, such as Carnap (1947), Church (1951), Kemeny (1956) and Ajdukiewicz (1967b). (On the other hand Russell overcame this confusion.) We shall do well to keep clear from this muddle while retaining instead the distinction (made by Bolzano, Frege, Russell and Carnap) between sign and construct as well as Frege's notion of sense (*Sinn*) – which, unfortunately, he did not elaborate on. (Recall Ch. 4.)

The third view, naive (or vulgar) materialism, is best represented by Buridan, Hobbes, Hilbert, Tarski, and Quine. Like naive idealism it is simple, clear, consistent – and mistaken. A sentence, *qua* string of sounds or ink marks, is an object for physical science (e.g., acoustics or chemistry). A physical object becomes a linguistic object the minute it is regarded as a medium for expressing something. As the astronauts have discovered, in an uninhabited world there are no more sentences than there are propositions: at most there may be inscriptions – the bones of dead sentences. Further, sentences can be well formed or ill formed. If the former then they may but need not express any propositions. Thus 'Aleph zero took a bath' is a sentence that stands for no proposition: this is why it is devoid of significance. (If sentences were assigned truth values then, to make room for nonsignificant or nonsensical sentences, we should adopt some system of three valued logic – which, to say the least, would be unpractical.) Finally a metaphysical point: to attribute to strings of marks or sounds any semantic properties, such as meaning and truth, is to indulge in hylemorphism *à la* Plato or Husserl. Following Leibniz (1703) we assume then that propositions, not sentences, are the direct bearers of truth: sentences can only be vicariously true.

We are thus left with the fourth view, conceptualist materialism – a sort of materialist version of conventionalism. It can be summarized as follows. Propositions are a kind of construct and as such they are useful fictions: we do not claim that they exist by themselves but only that it is often convenient (for example in mathematics but not in metaphysics) to feign or pretend that they do. We do not assert that the Pythagorean theorem exists anywhere except in the world of phantasy called 'mathematics', a world that will go down with the last mathematician. And we do not pretend that unknown propositions exist but find it advantageous to proceed *as if* all the logical consequences of a proposition, both the few known ones and the infinitely many unknown ones, did exist in some conceptual context. (After all this is what the most fervent nominalist does when he equates an axiomatizable theory with the set of consequences of the axioms of the theory.) In this way we keep the heuristic and metatheoretical advantages of Platonism, which allows us to handle infinite sets of statements, only a tiny fraction of which will ever be formulated, even less justified. But we avoid the weird metaphysical hypothesis that every possible proposition actually exists in a ghostly Realm of Ideas.

Our view can be clarified by contrasting it to the prevailing doctrine of truth, namely Platonism. To a Platonist such as Frege every proposition exists from all eternity and has a truth value even if we do not know which one. This is one reason for defining a proposition as anything that is either true or false. In our semantics propositions are not so defined: they are characterized instead together with predicates, namely thus: a predicate is a function from objects of some kind to propositions. (Cf. Ch. 1, Sec. 1.3.) Once we have formed a proposition we may *find out* its meaning and *assign* it a truth value. Meanings are discovered by investigating both the reference and the context (e.g. theory) in which the propositions dwell. In any case propositions are born with a fixed meaning. There is no such thing as a meaningless proposition (in contrast to a sentence devoid of significance). True, it often takes a lot of work to discover the full meaning of a proposition: it requires displaying the full context and the logical relations in it.

On the other hand propositions are not born with a truth value: this one is assigned, if at all, once the proposition has been formulated. There are propositions that have not yet been assigned a truth value, e.g. because we have been unable to prove (or disprove) them or confirm (or infirm) them empirically. Surely, if p happens to be a theorem entailed by premises which have already been evaluated, then we *discover* the truth value of p by proving p. But not all propositions are provable. Axioms are not, nor are data. And if a proposition is not provable then there is no truth value to be discovered. If p is not a theorem then either we assign p a truth value (on some ground or just for the sake of argument) or we don't. If we do, our assignment may be criticized and exchanged for another: we shall say that the truth value of p in the light of evidence e (empirical or theoretical) is v, only to point out that a different body e' of evidence might suggest a different truth value v'. And if we refrain from assigning p a truth value, e.g. for want of means or of interest, then p remains a proposition though one without a truth value. Which ruins the Platonic conception of propositions.

1.2. *Truth Values: Acquired*

Being constructs, propositions are *designated* (or expressed) by linguistic objects of a certain category, namely sentences. And, being constructs, propositions *have* a sense and a reference – the latter either actual or hypothetical. In addition to having a sense and a reference, some (actually most) propositions *can be assigned* a truth value. And a few such potentially true or false propositions are *actually assigned* a definite truth value – a value that need not be unalterable. To put it negatively: Not every proposition can be assigned a truth value, nor is every truth value assignment final.

To be more explicit, the following kinds of proposition are assigned no truth value at least within certain contexts:

(i) the propositions that are not decidable in (with the sole resources of) a given theory have no truth value in this theory;

(ii) the propositions containing empty descriptions, such as "the perfect man", are neither true nor false – unless they assert or deny the existence of their imaginary referents;

(iii) the statements that are formulated, but are neither postulated nor proved nor confirmed nor made plausible, have no definite truth values. For example, ⌜$p \vee \neg p$⌝ in intuitionistic logic, and ⌜There are black holes⌝ in current astrophysics and cosmology.

Since not every statement can be assigned a truth value we shall construe the truth value assignment (function) $\mathscr{V}$ as a *partial* function, i.e. as a function on a proper subset S_D of the set S of all statements. (I.e., we adopt the truth value gap view.) We shall characterize $\mathscr{V}$ in Sec. 3. But before doing so we must discriminate among the various kinds of truth, for different kinds call for different value assignment procedures.

Figure 8.1 summarizes our views so far.

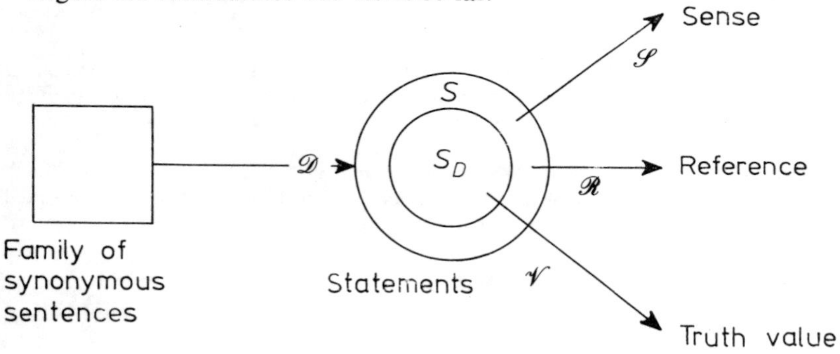

Fig. 8.1. Every set of synonymous sentences designates a statement in S. Every statement has a sense and a reference. Some statements are assigned a truth value.

1.3. *Quadruple Truth*

Consider the statements in Table 8.2, every one of which may be regarded as being true in its own context.

It would be nice if all these statements were true in the same manner, i.e., if a single concept of truth were applicable to them all. If this were the case then either the coherence or the correspondence doctrine might be able to span the whole set of statements. And if the former were the case, i.e., if every true statement were a *vérité de raison*, then model theory might suffice. That is, the semantics of mathematics would account not only for statements such as ⌜Formula ϕ is true in model M⌝ but also for propositions like ⌜The cost of living is rising steadily⌝. Unfortunately model theory does not help us with the second one nor with any statement in factual science, for the latter contains only fully interpreted formulas whereas model theory applies only if alternative interpretations (in mathematical structures – not with reference to the world) are possible. (Cf. Ch. 6, Sec. 2.) In particular, model theory is competent to

TABLE 8.2
Four kinds of truth: logical, mathematical, factual, and philosophical

Kind	Example
1 *Logical*	For any statement p, $\neg(p \& \neg p)$
2a Abstract	In a Boolean algebra, for any element x, $x \wedge \bar{x} = 0$
Mathematical	
2b "Concrete"	In an algebra of sets, for any set S, $S \cap \bar{S} = \emptyset$
3a Theoretical	Let p_0 be the probability of allele A in the first generation and μ the mutation rate from allele A to allele a. Then the probability of
Factual	the old allele A in the nth generation is $p_n = p_0(1-\mu)^n$
3b Empirical	Almost all the individuals in the 100th generation of the bacillus X were observed to be mutants in the trait Y that had been observed in the fraction p_0 of the original population
4a Semantical	Meaning precedes truth
Philosophical	
4b Metaphysical	Every thing changes

handle formulas that hold under every interpretation of the variables involved – i.e. in all models. (These are the tautologies.) In sum model theory, which formalizes and systematizes the coherence view of truth, is not universal: it does not even apply to the whole of mathematics.

On the other hand the correspondence theory of truth leaves us in the lurch with regard to logic and mathematics, which need not fit any facts in order to be valid. And it is unlikely that a third view could apply to all kinds of truth; at any rate no such all-encompassing theory seems to have been proposed. We had better recognize that the word 'truth' designates at least four different concepts, every one of which must be characterized in its own way: logical truth, mathematical truth, factual truth, and philosophical truth. Table 8.3 exhibits some of the salient peculiarities of mathematical truth and of factual truth. We shall take a

TABLE 8.3

Characteristics of three kinds of truth

Mathematical		Factual (e.g., biological)
Abstract (e.g., lattice theoretical)	"Concrete" (e.g., number theoretical)	
No formula has a truth value by itself, independent of an interpretation.	Most formulas have a truth value. Only the undecidable propositions lack a truth value (in the system in question).	Many formulas are assigned a truth value.
A relation between uninterpreted formulas and definite conceptual structures (models)	A relation between assumption and consequence.	A relation between statements and their referents.
Definable in terms of satisfaction (e.g., x satisfies ϕ in M) or of proof. Hence dispensable.	Definable in terms of provability in the case of syntactically complete theories. Hence dispensable in this case.	Not definable but characterizable. Indispensable.
Holds in a whole category of structures, e.g., the category of groups.	Holds only in a limited context, i.e., the theory of a model.	Holds only for a limited domain.
Can be total.	Can be total. Exception: approximation theory.	Can be total only in simple cases. Most theoretical factual statements are only partially true.
No decision procedure in most cases.	Usually decidable by proof or by counterexample.	No decision procedure. Only specific criteria for estimating truth values.
Truth valuation a purely rational procedure: experimenting infrequent and only heuristically valuable. Hence truth value independent of the world.		Truth valuation based partly on observation. Hence truth value contingent upon the world.

closer look at the former in Sec. 2.1, whereas the *vérités de fait* will be examined in Sec. 2.2.

2. Truth of Reason and Truth of Fact

2.1. *Truth of Reason*

A truth of reason is, of course, one that can be established by reason alone. A number of kinds of truth of reason have been distinguished, among them the following.

(i) *Dictionary truth* or *veritas ex vi terminorum* – e.g., a nominal definition.

(ii) *Truth by request* or postulation – e.g., a postulate in a mathematical theory.

(iii) *Truth by proof or deduction* – e.g., a theorem in a mathematical theory.

(iv) *Logical truth* or *veritas ex vi formarum* or tautology – e.g., any valid formula in a given system of logic.

(v) *Truth by exemplification* or satisfaction in a model.

The first two hardly deserve to be classed among truths: dictionary "truths" are mere conventions; and mathematical postulates are proposed because they summarize theories not because they are assumed to be true in themselves. (If abstract they may be true in or relative to some model, if "concrete" they generate truths by deduction.) The third, truth by deduction, is another case of abuse of the word 'truth': if its peculiarity is that it is deducible from a set of assumptions then the concept of truth is redundant here. Only the last two concepts of truth are legitimate, namely those of true under all interpretations (logical truth) and true under some interpretations (mathematical truth). They are the object of model theory (Tarski, 1954–55; Robinson, 1963; Bell and Slomson, 1969; Chang and Keisler, 1973). We shall presently take a look at the model theoretic concept of truth – but only to better clarify, by way of contrast, the notion of factual truth.

Consider the formula $\phi = \ulcorner x^2 + x = 0 \urcorner$. This semiabstract formula is *satisfied* by the number -1, i.e. it becomes a *true* formula when x is interpreted as -1. Equivalently: the valuation $x := -1$ satisfies ϕ or produces a true statement. Also: ϕ is *satisfiable* in the structure $\mathscr{Z} = \langle Z, +, -, 0 \rangle$, where Z is the set of integers. But the very same formula is also satisfied by the negative of any unit matrix I. Equivalently: ϕ is *satisfiable* in the ring $\mathcal{M}$ of square matrices. And so on and so forth. In symbols:

$$\models_{\mathscr{Z}} \phi(-1), \qquad \models_{\mathcal{M}} \phi(-I), \quad \text{etc.}$$

The notion of satisfaction can be extended to a set of formulas. For example, all of the axioms of the abstract theory of groups are satisfied by, or hold for, the integers. A proof that the structure $\mathscr{Z} = \langle Z, +, -, 0\rangle$ is or exemplifies a group $\langle S, \circ, ', e\rangle$ consists in showing that the valuation (interpretation)

$$S := Z, \qquad \circ := +, \qquad ' := -, \qquad e := 0$$

satisfies the axioms of group theory taken jointly. In other words, these abstract formulas (open formulas), when assigned the above interpretation, become formulas recognized as true in a specific (often familiar) field – in this case elementary number theory. The validity of the latter goes unquestioned in the proof. What one shows is that one conceptual structure (the abstract one) fits another conceptual structure (a specific or "concrete" one) – a clear case of *truth as coherence*.

This model theoretic concept of truth as satisfaction in a model serves to define the concept $\vdash$ of deducibility, namely thus. The formula ⌜If x and y are in L, then $x \wedge y \leqslant x \vee y$⌝ can be said to *follow* from the axioms of lattice theory because, in *every example* of a lattice, meets precede joins. On the other hand a special theorem for chains, such as ⌜For any x and y in L, $x \leqslant y$ or $y \leqslant x$⌝, does not follow from the axioms for L. In general: let A be an abstract axiom system and let ϕ be a formula containing only concepts occurring in A. Then $A \vdash \phi$ iff every model of A is a model of ϕ, i.e. $A \models \phi$. This relation between $\vdash$ and $\models$ is easily generalized to a whole set S of abstract formulas. Call $\mathcal{M}(S)$ the whole lot of structures in which the formulas in the set S are satisfiable. Then the theory of $\mathcal{M}(S)$ is the set $\mathscr{C}n(S)$ of logical consequences of S. The notion of truth by proof is thus reduced, in a sense, to that of truth by exemplification or truth in a model. That this does not eliminate the methodological differences in proof types and procedures is beside the point: semantics is not concerned with methodological problems.

We have recalled succinctly the model theoretic concept of truth only to bring out its differences with the concept of factual truth employed in factual science. The most outstanding differences are these:

(i) The objects said to be true in some (extensional) model, or under some interpretation, are *abstract formulas.* What satisfies or fails to satisfy an abstract formula is always some construct or other. Because snow is not a mathematical object, it makes no sense in mathematics to state that snow satisfies the semiabstract formula ⌜x is white⌝. Nor does it make sense in the context of factual (ordinary or scientific) knowledge, unless we are willing to assign a semantic property to a material thing.

(ii) An abstract formula, or propositional function, if satisfiable, and a fortiori if valid, is so *in one or more models.* By contrast, a factual statement need have no mathematical status at all: think of ⌜There are many beautiful children⌝. Moreover a factual statement, if true (to some extent), is true *of* the world not *in* a model. It would be ludicrous to write, say,

$\models_W$ Maxwell's electromagnetic equations, where W is the world.

if only because the real world is not a mathematically defined structure (an extensional model *à la* Tarski). The fashionable interpretation of models (or else of model sets in the sense of Hintikka (1969)) as possible worlds serves just one useful purpose – that of dressing Leibniz's notion of logical truth in modern garb. (To wit: p is logically true iff p holds under every interpretation of the nonlogical constants occurring in p, i.e. if p is satisfiable in every model – intuitively, if it holds in all conceivable, or conceptually possible, worlds.) This pseudo-ontological interpretation of model theory elucidates neither the concept of truth of fact nor that of real possibility. The former if only because truths of fact are more often than not partial, whereas a model set is constituted by formulas that are totally true under some interpretation. As to possibility, not any old model set describes a possible state of affairs or a possible course of events ("world"); only a set of law statements can do that. (The statements are at best approximately true when the possibilities are actualized. As long as the events are merely possible the corresponding statements do not form a model set.) Moreover, the translation of 'model' into 'possible world' has misled some philosophers into believing that, since logical truths hold (are valid) in every "possible world" (model or alternatively model set), and since the actual world is possible, logical truths must hold of reality, so that logic constitutes the basic ontology (Scholz, 1941) rather than a metaphysically neutral discipline. (More in Bunge, 1974a.)

To conclude. The model theoretic concept of truth elucidates the previously imprecise *coherence* theory of truth – contrary to Tarski's intention that it should formalize the correspondence theory. It also elucidates the notion of truth by proof. Consequently that semantic concept is of importance to pure mathematics. But it is irrelevant to factual knowledge, where we have not only coherence (or the fitting of constructs to one another) but also external reference.

Tarski's semantic conception of truth, in its mature (model theoretic) shape, revolutionized metamathematics. Moreover it drew the attention of some philosophers to a long neglected (or rather repressed) yet central problem of philosophy. But at the same time it persuaded some of the very best among them (notably Carnap, Popper, and Quine) that there was nothing problematic left about truth of fact. Yet Tarski's theory did not even attempt to grapple with the latter – so much so that all the technical work in model theory, starting with Tarski's pioneer papers (1936, 1954, 1955), is exclusively concerned with theories in pure mathematics – moreover with "formalized languages", i.e. abstract theories. Only an idealist could take model theory to be applicable to factual knowledge as well – since for him the world is a realization of an abstract idea. Let us then part company with model theory and take a fresh look at factual truth. (We shall return to Tarski's theory of truth in Sec. 2.4.)

2.2. *Truth of Fact: The Synthetic View*

A factual statement is one involving at least one factual predicate (see Ch. 6, Sec. 4.1). And factual truth (or falsity) is predicable of a factual statement in relation to some domain of facts and irrespective of its mathematical status.

Since the coherence theory of truth is not concerned with factual reference, it would seem that we must turn to the correspondence view of truth. According to the latter a statement is true if it fits the facts. Unfortunately the nature of this fitting has never been clarified: most of the time it is left in the twilight of metaphor, occasionally it is explained as isomorphism. Let us brush metaphors aside: they constitute no theory. Nor does the isomorphism thesis constitute a theory. To begin with isomorphism can only be defined between mathematically well defined structures – and reality is no such structure. Secondly, even the intuitive notion of isomorphism fails us here as shown by the facts that (*a*) every scientific theory contains constructs with no real counterpart and serving at the most as computational devices, and (*b*) every piece of reality ends up by showing traits that had not been contemplated by any theory. If there is correspondence between theory and facts it must be global not point by point. But it is doubtful whether this global correspondence can suffice to characterize factual truth. Indeed, consider the following cases, every one of which constitutes a counterexample to the (unformulated) correspondence theory of truth:

(i) ⌜There are no ghosts⌝ is factually true precisely because there is no such thing as a ghost.

(ii) Continuum mechanics is nearly true of most bodies – which are known not to be continuous.

(iii) The predictions and retrodictions calculated on theories assumed to be true refer to possible facts.

The reason for the failure to take the correspondence doctrine of truth beyond the metaphor phase and to have it surmount the indicated difficulties is both simple and radical, namely this. Statements cannot be compared or confronted with facts. Statements can only be confronted with other statements, and facts with other facts only. The expression 'Confronting a proposition with a fact' should be taken as short for 'Confronting a judgment (=brain process) with another fact' or, simpler, 'Thinking of a fact'. What holds for confrontation holds for fitting. A statement does not fit facts the way clothes fit people: it can only "fit" another statement, or "agree" with it barring certain details. In any case semantics is not equipped to investigate the mental processes of confrontation and adaptation of ideas to facts: it can only handle the confrontation among statements.

The following kinds of *inter-statement confrontation* are of particular interest to factual science:

$t-t'$ theoretical statements *vs.* theoretical statements
$t-e$ theoretical statements *vs.* empirical statements
$e-e'$ empirical statements *vs.* empirical statements.

Examples.

$t-t'$ The probability of a radiative transition of an atom from one energy level to another, calculated in nonrelativistic quantum theory, *vs.* the same probability calculated in relativistic quantum theory.

$t-e$ The probability of a radiative transition of an atom from one energy level to another, calculated in some theory, *vs.* the intensity of the radiation of a collection of atoms of the same kind, measured with a given equipment and technique.

$e-e'$ The intensity of the total radiation of a collection of atoms measured with a certain equipment and technique, *vs.* the values of the same magnitude measured with a different equipment and technique.

Any two statements can be put side by side. But if the aim is to estimate truth values then only statements with a *common meaning* (not necessarily identical) should be paired. In other words, for two statements to compete they must share some of their sense and some of their referents. (See Figure 8.2.) Thus while it may be fruitful to compare the values of a reaction time obtained by different methods, it would be foolish to compare either with the international price of sugar.

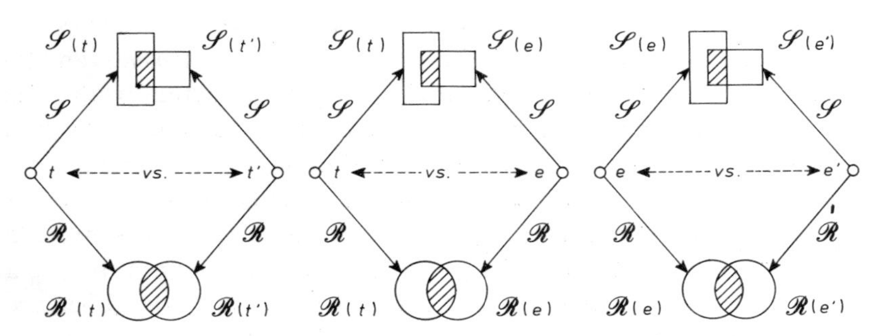

Fig. 8.2. Confronting scientific statements. Condition: both the senses and the reference classes must have nonempty intersections. Warning: $\mathscr{S}(e)$ and $\mathscr{S}(e')$ are rather vague.

The common meaning condition is most readily satisfied when the two statements are either theoretical or empirical. Difficulties arise when one of them is theoretical and the other empirical. And if the latter happens to be experiential (a sense datum), rather than observational or experimental, then there may be no point in confronting it with a physical object statement. For example, ⌜I am feeling hot⌝ differs both in sense and in referent from ⌜The present air temperature is 40 °C⌝. Hence neither can refute the other. In general, sense data are of little use in science precisely for this reason. Before a datum can become an item of evidence it must be depersonalized. But this is not enough: the datum must also be couched in the language of the theory it is supposed to be confronted with. For example, pointer readings must be interpreted as values of magnitudes handled by some theory. And before the latter can be confronted with any data it must be adjoined to special assumptions relevant to the situation that is being handled, as well as subjected to certain purely mathematical operations such as integration or substitution of reference frame. In short, the raw data are not comparable to the pure theory: the former must be lifted up to a theoretical level which is in turn lower than the axiom level. All this is a matter for the special sciences as well as for the general methodology of science (see e.g. Bunge, 1967a, Ch. 15, and 1973b, Ch. 10). It is also a matter for methodology to decide the conditions under which two comparable statements *agree* with each other, perhaps to within the preassigned experimental error. (See the next section.)

What is of overriding interest to semantics is (*a*) that the theory-fact confrontation boils down to confronting two sets of *statements*, and (*b*) that a theoretical statement is declared to be true if it "agrees" with some other *statements* (some of them empirical and others theoretical). It would seem that the coherence theory of truth does hold for factual science after all. It does but only in part: although a factual theory is true just in case it "agrees" with another set of propositions, all the parties involved – those in the dock as well as those in the jury box – happen to have a factual reference. We arrive thus at a sort of synthesis of the coherence and the correspondence views of factual truth – whence the name *synthetic* view. The coherence consists in inter-propositional agreement, the correspondence in factual reference. And the correspondence or degree of adequacy is tested by coherence: the latter supplies the criterion of truth not a definition of it. (Cf. Rescher, 1973.)

Another, no less important feature of factual truth of interest to semantics is this: it is *partial* more often than total – i.e. it comes in degrees. Figure 8.3 illustrates this contention (not contended by scientists though

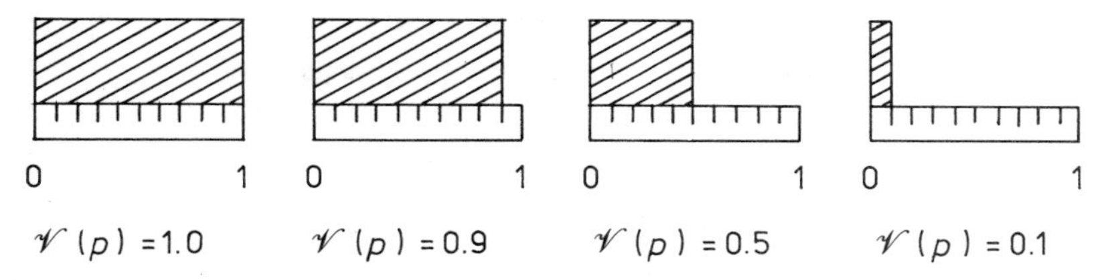

Fig. 8.3. Outcomes of a truth valuation of $p = \ulcorner$The box is 1 cm long$\urcorner$. (*a*) Complete truth; (*b*) approximate truth; (*c*) half truth; (*d*) near falsity.

brushed aside by logicians). The truth valuation function $\mathscr{V}$ occurring in the caption will be defined by Criterion 8.1 in Sec. 2.3. In the case of Figure 8.3 it is assumed that the ruler indicates its own actual length as well as that of the box. This hypothesis is the basis for assigning p a truth value.

2.3. *Truth Values: Conditional*

Although methodology is no substitute for semantics it may supply useful clues for investigating the semantics of science. In particular, if we wish to find out what factual truth is we shall find it useful to become acquainted with the way truth values are assigned in science. A very quick review will suffice for our purposes.

Whenever two statements, s and s', are confronted with a view to evaluating either of them, the following situations can arise:

(1) *Neither statement is taken for granted* (presupposed or assumed).

(*A*) *The two statements turn out to agree.* Unless there is a logical relation between them, both statements are assigned the same truth value: $\mathscr{V}(s) = \mathscr{V}(s')$. If s implies s' then $\mathscr{V}(s) \leqslant \mathscr{V}(s')$. In either case there is mutual confirmation but no independent assignment of truth values: for all we know the two statements could be equally false. (Moral: Confirmation, though necessary, is insufficient.)

(*B*) *The statements disagree.* They are assigned different truth values: $\mathscr{V}(s) \neq \mathscr{V}(s')$ – but we still do not know how much each is worth. (Moral: Refutation too is insufficient.)

Upshot: If neither of the two statements is assumed, at least provisionally, the other cannot be assigned a truth value either.

(2) *One of the statements, say s, is assumed* (unquestioned in the given investigation).

(A) *The two statements turn out to agree.* Both are assigned the same truth value, namely one or nearly so. Possible cases:

(*a*) Both statements are theoretical. Theory confirms theory.
(*b*) s is theoretical, s' experimental. Theory endorses experiment.
(*c*) s is experimental, s' theoretical. Experiment supports theory.
(*d*) Both statements are experimental. Experiment supports experiment.

(*B*) *The statements disagree.* The statement under examination is assigned a truth value below unity: $\mathscr{V}(s') < \mathscr{V}(s) = 1$. Possible cases:

(*a*) Both statements are theoretical. Theory infirms theory.
(*b*) s is theoretical, s' experimental. Theory indicts experiment.
(*c*) s is experimental, s' theoretical. Experiment infirms theory.
(*d*) Both statements are experimental. Experiment indicts experiment.

(*C*) *The statements agree in a certain region R but disagree in $\bar{R}$:* see Figure 8.4. The statement under examination is assigned a truth value dependent on the point x of the explored area: $\mathscr{V}(s', x) = 1 - \varepsilon(x)$, where $\varepsilon(x)$ is the discrepancy between s' and the base line s at the point x. Possible cases:

(*a*) Both statements are theoretical. Theory confirms theory in region R, disconfirms it in region $\bar{R}$.
(*b*) s is theoretical, s' experimental. Theory confirms experiment in region R, indicts it in $\bar{R}$.
(*c*) s is experimental, s' theoretical. Experiment confirms theory in region R, infirms it in $\bar{R}$.

(*d*) Both statements are experimental. Experiment strengthens experiment in region R, weakens it in $\bar{R}$.

Upshot: If one of the two statements is assumed, if only provisionally, the truth value of the other can be estimated or at least bounded.

We conclude that factual truth values are *conditional* or *relative* not

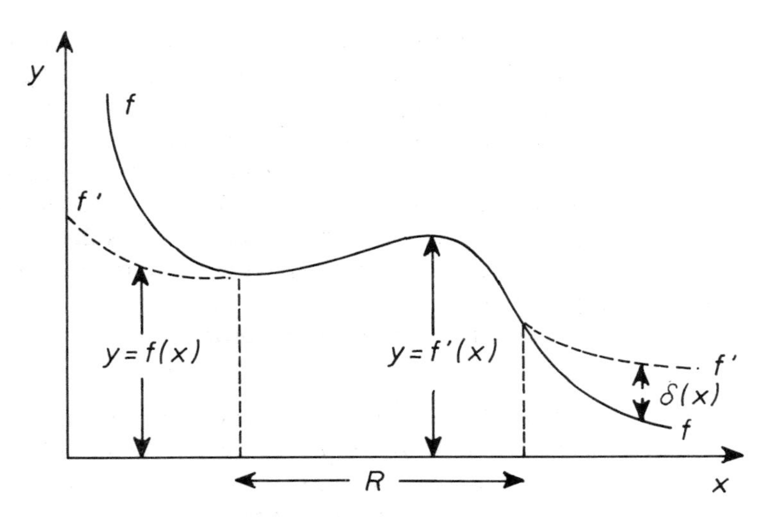

Fig. 8.4. A common situation in science: two statements, $s=(y=f(x))$ and $s'=(y=f'(x))$, agree in region R but disagree elsewhere. The relative discrepancy or error $\varepsilon(x)$ depends on the point x of the explored area and is proportional to the absolute discrepancy δ (x).

absolute. Hence strictly speaking we should always write '$\mathscr{V}(s' \mid s)$', read 'the truth value of s' given, or assuming, s', instead of '$\mathscr{V}(s')$'. And, since there are only relative factual truth values in science, there would be no point in trying to analyze $V(s' \mid s)$ into the absolute truth values $V(s')$ and $V(s)$ similarly to the way conditional probabilities are analyzable into absolute ones. Which suggests that truth values cannot be probabilities. (More on this in Sec. 5.1.)

In general we shall not encumber our formulas with a symbol for the statement serving as a yardstick for the estimation of the truth value of another statement. But it will be convenient, nay indispensable, to indicate the baseline, in some way or other, whenever actual truth value assignments are made. A typical case in science is the evaluation, by alternative methods, of some magnitude M concerning an object b – for example the relative dilation of a metal bar. Suppose we have the following confrontation:

$$s = \ulcorner M(b) = m \urcorner \quad vs. \quad s' = \ulcorner M(b) = m' \urcorner,$$

where m and m' are rival numerical values obtained for the M-ness of b. (We are brushing aside, for being irrelevant to our purpose, the M-units as well as the experimental error.) The absolute value of the numerical discrepancy between the two statements is $|m' - m|$: this is the *error* made in accepting s' instead of s. If the error is small, then the truth value of s' will be close to that of s, i.e. near unity. But as the error increases the truth value of s' approaches zero. In general this error will depend on the referent b and its condition; hence the relative truth values themselves will exhibit this dependence.

The preceding considerations suggest the following

CRITERION 8.1 Let M and M' be two comparable functional representations of a given property of an object b, and let $s = \ulcorner M(b) = m \urcorner$ and $s' = \ulcorner M'(b) = m' \urcorner$ be estimates of M and M' respectively at b. Then the *relative truth value* of s' given (assuming) s equals

$$\mathscr{V}(s' \mid s) = 1 - \left| \frac{m - m'}{\operatorname{Max}\{m, m'\}} \right|$$

If the error is small, $\mathscr{V}(s' \mid s)$ comes close to 1; if the error is large, the truth value is near 0. *Example:* Let us compare the truth value of Boyle's law (B) relative to the law of Boyle and Mariotte (M), i.e., making the pretense that the latter is true. Since the statements are

$$M = \ulcorner p = aT/v \urcorner, \qquad B = \ulcorner p = b/v \urcorner,$$

we have

$$\mathscr{V}(B \mid M) = 1 - \frac{aT/v - b/v}{\operatorname{Max}\left\{\frac{aT}{v}, \frac{b}{v}\right\}} = 1 - \left| \frac{aT - b}{\operatorname{Max}\{aT, b\}} \right| .$$

Both for very low and very high absolute temperatures $\mathscr{V}(B \mid M)$ is near zero; it comes near unity for medium temperatures and certain gases. That truth values should depend on the kind of material and on the range of the physical variables involved may sound outrageous to both pure logicians and Platonists. But it is no news to a scientist, who is used to thinking of (factual) truth as holding to some extent *of*, or *with regard to*, definite external referents in definite conditions. This is precisely the peculiarity of factual truth: that it concerns facts. For this reason, as will be seen in the next subsection, the semanticist has no say in setting up truth conditions for factual statements.

Criterion 1 applies, in particular, to the comparisons of theoretical or calculated values with experimental values. In this case we have pairs of propositions such as these:

Theory $M(b)=m_t$, with m_t a knowable number

Experiment $M'(b)=m_e \pm \varepsilon_e$, with m_e and ε_e knowable numbers (m_e is the measured value and ε_e the random error characteristic of the particular run of measurements leading to m_e).

Unless the possible values of M are well spaced, the chances are that the theoretical value m_t and the central experimental value m_e differ from one another – and both from the real value. We need then definite criteria allowing us to make definite choices. One such criterion, actually used if not formulated explicitly in the sciences, is

CRITERION 8.2 Let m_t and m_e be a theoretical and an experimental estimate, respectively, of a magnitude M representing a property of a thing b. Then

(i) $\ulcorner M(b)=m_t \urcorner$ and $\ulcorner M(b)=m_e \urcorner$ are equivalent, to within the error ε_e, iff $|m_t - m_e| < \varepsilon_e$;

(ii) the "true" (or "real") value of M for b is close to m_t iff (a) m_t coincides with m_e to within ε_e (i.e. if (i) holds) and (b) $\varepsilon_e / m_e \ll 1$.

2.4. *Truth Conditions*

A truth condition (or criterion) for a set S of formulas is a metastatement stipulating the conditions under which the members of S are (fully) true. Example:

If x is a real number then $\ulcorner x^2 > x \urcorner$ is *true* iff $|x| > 1$ or, equivalently,

$\ulcorner x^2 > x \urcorner$ is *satisfied* by all real numbers x such that $|x| > 1$.

Tarski's theory of truth (Tarski, 1936), based on the concept of satisfaction, provides a general truth condition for all the *abstract* formulas of logic and mathematics and it has become part and parcel of model theory (see e.g. Hermes, 1963). Some philosophers believe that the same trick works for all statements, in particular for factual statements. (And it certainly should if, as Tarski and Quine hold, there is no radical difference between the factual and the formal nor, a fortiori, between the synthetic and the analytic.) Thus Carnap illustrates the theory by contriving a toy semantic system for an object language with 7 specific (extralogical) signs: the individual constants x_1, x_2 and x_3, the predicate letters P_1 and P_2, and the left and right parentheses (Carnap, 1942, pp. 22ff.). The atomic sentences are all of the form "$P_i(x_j)$" with $i = 1, 2$ and $j = 1, 2, 3$. There is one designation rule for every specific sign, in particular the following:

'x_1' designates Chicago.

'P_1' designates the property of being large.

The truth condition for the sentences of the microlanguage is

The sentence '$P_i(x_j)$' is true iff the designatum of x_j has the designatum of P_i.

An instance of this truth condition is, of course,

'$P_1(x_1)$' is true iff Chicago is large.

It would be nice if Tarski's theory of truth were a stretch theory fitting all kinds of statements, whether formal or factual. But it is not. First, in the body of factual knowledge there are no *uninterpreted* formulas, such as '$P_i(x_j)$', containing predicates with no fixed sense. Hence the possibility of having factual formulas satisfied in alternative models is not forthcoming – whence model theory is irrelevant to our concern.

Second, to say that Chicago *satisfies* the open formula $\ulcorner x$ is large$\urcorner$ involves assigning a nonphysical (namely a semantic) property to a physical entity – which is unacceptable to a non-Platonist. Things are the other way around: $\ulcorner x$ is large$\urcorner$ has the (semantic) property that, when the dummy x is replaced by a name for Chicago it becomes a true proposition.

Third, there is no *universal* truth condition for all of the (interpreted) formulas with a factual reference, from $\ulcorner$Chicago is large$\urcorner$ to the sophisticated equations of mathematical biophysics. There are at most regional truth conditions such as

> ⌜The city of x is large⌝ is *true* iff the population of x is over one million people.

In this simple case the truth condition proves to be a trivial convention of demography – one that is neither portable nor eternal nor stipulated by semantics. In other cases, *if* there are truth conditions they are more complex. And in any event they are laid down (and upset) by the competent discipline. It should be obvious why this is so: a factual truth condition depends on the specific meaning (sense and reference) of the corresponding statement as well as on the possible procedures for putting it to empirical tests. Consequently factual truth conditions cannot be cooked up by semantics.

Fourth, in factual science there are hardly any truth conditions having the neat form of biconditionals of the form: ⌜*A*⌝ *is true iff B*, a generalization of Tarski's principle: ⌜*A*⌝ *is true iff A*. The most we get in factual science are conditionals such as ⌜If the theory (or hypothesis) T is true then the effect e is observable⌝. (Or, equivalently, the corresponding counterfactual sentence 'If T were true then e would be observable.') But such conditionals work as clues for guessing at truth values not as criteria for assigning definite truth values. Indeed, the validation of the consequent e confirms the antecedent T without verifying it: in principle infinitely many alternative constructs T', T'',... might be substituted for T. It is only a whole lot of confirmations together with the compatibility of T with previously corroborated theories that allows one to (tentatively) assign (approximate) truth values to T. In sum, the normal situation in factual science is the lack of neat truth conditions and the presence of entire batteries of tests for (partial) truth (Bunge, 1967a, Vol. II, Ch. 15). And such assignments of degrees of truth are rarely final. (See Sec. 4.4.)

Fifth and lastly, since factual truth is rarely total, the truth conditions found in the usual treatments of mathematical logic are often inapplicable to it. A half truth such as ⌜Aristotle was a Chaldean philosopher⌝ fits neither of the standard truth conditions ⌜*A & B is true iff A is true and B is true*⌝, and ⌜$\neg$*A is true iff A is not true*⌝. (Roughly, since one of the conjuncts is totally true and the other totally false, A & B is worth $\frac{1}{2}$ and so is its negate.) Hence logic, indispensable as it is for the control of inference, is utterly impotent to guide our assignment of factual truth values.

Instead of uniform and unchanging truth conditions, what we do meet in factual science are three kinds of conditions, all of them regional or subject matter dependent. We shall introduce them by way of example. Consider Galilei's law of falling bodies. There is, first, the *applicability condition* indicating the referents and the state they are in – for example a body immersed in a homogeneous gravitational field and in free fall in it. This condition occurs as the antecedent of the law statement: ⌜If a body falls freely in a constant gravitational field in vacuum, then G⌝, where $G = ⌜v(t) = gt + v_0⌝$. The antecedent is not necessary for the truth of G but for its applicability or relevance: if the condition fails to hold the conditional is true but pointless. And the applicability criterion is *intra*theoretical: it owes nothing to test conditions. Rather on the contrary, testing presupposes that the applicability condition holds.

The second condition may be called an *ontological truth condition*, for it points to those referents that behave actually as indicated by the statement in question. In other words, the ontological truth condition for a statement is learned from experience and it exhibits the extension of the given formula. For example, the extension of G above is the collection of triples ⟨middle sized body, weak constant gravitational field, short time of fall⟩. This condition is *extra*theoretical and a posteriori. However, it is not theory free, for it sums up the outcome of tests conducted in the light of other (instrumental) theories and it evaluates the performance of the statement concerned in relation to (actual or possible) competitors.

A third and last metastatement of interest specifies the methodological conditions under which the given statement may be regarded as approximately true. It is, like the former, extratheoretical but, instead of specifying the kind of thing for which the statement holds (approximately), it concerns the particular empirical techniques employed in assaying it: this condition may therefore be called an *epistemic truth condition.* One such condition, in our example, would be this: ⌜G holds to within 1% for steel balls, in air, at sea level, and for distances of the order of 10 m, when tested with a sports chronometer and a commercial tape⌝.

All three conditions are special or subject matter dependent. Consequently they cannot be laid down by semantics. The semanticist may of course study them on condition that he leaves the methodological aspect to methodology and concentrates on the alethic aspect. But for this he needs a theory of degrees of factual truth, one elucidating the intuitive notion employed in factual science. We turn to one such theory.

3. Degrees of Truth

3.1. *The Problem and How to Fail to Solve It*

The notion of degree of truth, and the related notion of approximate truth, are employed everywhere in applied mathematics and in factual science. Examples: (*a*) most values of nonalgebraic functions, such as *log* and *sin*, are known only approximately; (*b*) all nontrivial measurement results are approximate; (*c*) all theoretical statements are, at best, good approximations – and we hope to be able to improve on them. Provided certain assumptions are made, i.e., provided some statements are taken to be wholly true, one can often estimate the goodness of the approximation, i.e., the deviation from the truth. In particular, one may compute (*a*) differences between values given by rival theories, (*b*) discrepancies between theoretical and empirical values, and (*c*) random errors of measurement. Mathematical statistics, by elucidating several concepts of error and computing their probabilities, helps us to estimate the deviation from the truth and, in this way, the degree of truth. On the other hand neither the theory of probability nor mathematical statistics have rules for assigning probabilities to either hypotheses or data: this assignment has always been a philosophers' game.

Since applied mathematics and factual science are shot through with the concept of approximate truth, it is a task for the semantics of science to elucidate this concept, i.e., to propose theories of degrees of truth matching scientific practice. There have been several attempts to do so. They fall into four main classes: (i) many valued logics (e.g. Moisil, 1972); (ii) the semantic interpretation of probability, i.e. the equating of probabilities and degrees of truth (e.g., Łukasiewicz, 1913; Reichenbach, 1949); (iii) Popper's theory of verisimilitude (Popper, 1963b), and (iv) the author's former theory of partial truth (Bunge, 1963a). None of these can be counted as successful, although the last may come close to providing a realistic explication of the concept of degree of truth, if only because it does not involve the mythical concept of probability of a statement.

Knowing why those attempts failed may help in avoiding similar mistakes. The reasons for those failures are, in a nutshell, the following. The systems of many valued logic may be mathematically interesting and they have exerted a liberating influence by showing that ordinary logic is neither logically necessary nor psychologically compelling. But none of them has attained the maturity needed to cope with actual deductive inference as practised in mathematics. None of them has ever been used in mathematics or in science, where ordinary logic is perfectly adequate. Moreover, none of them is likely to be used. First, because changing the logic in any one field of research would require changing it in all related fields: the revolution would have to spread all over mathematics and science in order to allow for contacts among theories. Second, it would be imprudent to relax the standards of criticism (Popper, 1970). Third, the main rationale for many valued logics is a mistake, namely the belief that logic should be a theory of truth. Logic must be regarded instead, in line with the Aristotelian tradition, as a theory of deduction not as a theory of truth. We should then be able to keep the calculi of ordinary logic even if we intend to adopt a many valued theory of truth. This policy will be implemented in the next subsection.

As to the probability theories of truth, they all rest on the mistaken assumption that there *are* ways of assigning probabilities to statements. In point of fact, whereas scientists often succeed in calculating and measuring probabilities of certain *facts*, e.g., events, no one has ever proposed a general procedure (other than arbitrary fiat or else betting) for assigning any probability values to *statements*. Mathematics would have nothing against it, for the set of statements qualifies as the support of a probability measure. But it so happens that there *are* no rules for assigning numerical values to such probabilities, as a consequence of which no one has ever succeeded in estimating the probability of any given factual statement. This difficulty alone disqualifies all the theories of truth degrees based on probability, whether they equate truth with probability or with some function of the latter. (More on this in Sec. 4.2.) On the other hand there are more or less definite rules for estimating relative degrees of truth, as we saw in Sec. 2.3.

Finally the author's previous theory of partial truth does not hinge on the concept of probability and it leaves ordinary logic intact. Besides, it incorporates the notion of discrepancy or error as it occurs in the theory of errors. However, it has some serious defects pointed out by the author himself (1963a) and by some readers. For one thing it overrates confirmation. For another its truth function is discontinuous. Thirdly, its multiplication theorem (or rather axiom) is so complicated that it is nearly impossible to compute by hand the truth value of a conjunction with a reasonable number of conjuncts. We shall presently expound an alternative theory of degrees of truth that shares the virtues but not the defects of the former theory.

The new theory to be presented is based explicitly on an *alethically neutral conception of logic* – one that can be attached to any number of alternative theories of truth. This construal of logic is nothing but an explicitation of Bolzano's remark that we should distinguish a proposition from the statement (actually a metastatement) that the proposition is true. Briefly, $p \neq \ulcorner \mathscr{V}(p) = 1 \urcorner$. One advantage of the neutral conception of (formal) logic is that it allows us to talk about partial truth and moreover to hitch any theory of partial truth to the calculi of ordinary logic. Scientists, though perhaps not formal logicians, will appreciate this advantage. A second advantage of this truth-free conception of logic is that it permits us to employ the dialectical method – in Parmenides' not in Hegel's sense. Indeed, the dialectical method, universally employed in mathematics and in science, consists roughly in exploring the consequences of an assumption *before* evaluating it and *in order to* evaluate it. To carry out this method we must assume that the propositions under examination obey the laws of logic whether or not they turn out to be true. The most we might need is the pretense that statements are true or false whether we know it or not. But we take this as a make-believe useful for heuristic purposes: we assume explicitly only that (some) statements *can be* assigned truth values, or rather truth degrees, not that they are born with an intrinsic and everlasting truth value.

Compare this formal or alethically neutral conception of logic with the alethic interpretations. Among these, the standard view is the model theoretic or referential one, which employs the notions of satisfaction and of (formal) truth, e.g., in setting up truth conditions such as truth tables. An alternative view that is receiving some attention these days is the so called substitution interpretation (Barcan Marcus, 1962). According to this view '$(\exists x)\,Px$' should be interpreted as "Some substitution instance of Px is true" – and similarly for universal statements. Either view may be adequate for logic but, by the same token, it does not seem suitable for the applications of logic. For one thing they render the dialectical method inapplicable, insofar as they demand that every proposition be true or false from birth, whether we know it or not: i.e. it does not make room for tentative truth value assignments. For another, the alethic conceptions of logic employ a single concept of truth and moreover the one of total truth. If applied to factual science, where evidence is never complete and final, the alethic conceptions of logic may lead to the aberration of thinking that, precisely for those shortcomings, factual science need not abide by ordinary (classical) logic. (For a candid statement of this dangerous thesis see Birkhoff, 1961, Ch. XII.)

But enough of criticism: let us see how we can merge two valued logic with the idea that truth, unless it is formal, comes in any number of shades.

3.2. *Axioms*

Our theory of degrees of truth will treat complete truth and complete falsity as the two end points of a whole gamut of truth values, while retaining all the algebraic features of ordinary logic, for the latter is built into mathematics and science. In other words we shall (*a*) assume logic to be given in a purely syntactic fashion rather than with the help of (formal) truth values, and (*b*) join to it a valuation function $\mathscr{V}$ with values in a numerical interval, say the unit interval of the real line.

One way of implementing this programme is as follows. Consider the set S of all statements in a given field of inquiry, such as a scientific theory. Group all the statements in S that are logically equivalent to each other. That is, form the equivalence class $[s]$ of every statement s in S under the relation of logical equivalence: $[s]=\{s' \in S \mid \ulcorner s' \Leftrightarrow s \urcorner \text{ is a tautology}\}$. Call $[S]$ the set of all such equivalence classes, i.e. the quotient of S by the relation of tautological equivalence. It is well known that $[S]$ has the lattice structure. (This holds also for any extension of a given S but it need not hold for the union of arbitrary sets of statements, as they may be mutually incompatible. Thus the union of classical mechanics and quantum mechanics does not possess the lattice structure.) Moreover $[S]$ is a complete complemented and distributive lattice with null and unit elements – in short a Boolean algebra. The Boolean operations on the set $[S]$ of equivalence classes are defined in logical terms as follows: for any p, q, r in S,

$$\overline{[q]}=[p] \text{ iff } \ulcorner p\Leftrightarrow q\urcorner \text{ is a tautology}$$
$$[q]\cup[r]=[p] \text{ iff } \ulcorner p\Leftrightarrow q\vee r\urcorner \text{ is a tautology}$$
$$[q]\cap[r]=[p] \text{ iff } \ulcorner p\Leftrightarrow q\,\&\,r\urcorner \text{ is a tautology.}$$

Likewise the least element $\square$ and the last element $\mathord{▯}$ of the Boolean algebra of the equivalence classes of statements are defined by

$$\square=\{p\in S \mid p\Leftrightarrow q\,\&\,\neg q \text{ and } q\in S\}$$
$$\mathord{▯}=\{p\in S \mid p\Leftrightarrow q\vee\neg q \text{ and } q\in S\}.$$

(It is true that the algebra of quantification is far more complicated than this. However we need not go into it if our purpose is restricted to computing the truth value of truth functional compounds in terms of the truth values of their components. Nor need we stop at the possible objection that the formulas of quantum mechanics fail to constitute a distributive lattice. This opinion is false: suffice it to recall that the quantum theories, just as any other scientific theories, involve only classical mathematics, which has classical logic in its bones. Cf. Bunge (1967b) and Fine (1968).)

We now assume that there is a real valued function $\mathscr{V}$, defined on a certain subset S_D of S, such that

$$\text{for every } p \text{ and } q \text{ in } S_D,\ \mathscr{V}(p\,\&\,q)+\mathscr{V}(p\vee q)=\mathscr{V}(p)+\mathscr{V}(q).$$

The statements in the complement $S-S_D$ have no truth value because none can be assigned to them. In this residual subset we find the statements in S that cannot be proved with the sole resources of S, as well as the statements containing empty descriptions, such as "The perfect man does not exist" and "The massless body is not affected by gravity". (See Ch. 9, Sec. 2.) In addition to the above conditions on the partial function $\mathscr{V}$ we shall postulate that $\mathscr{V}$ assigns contradictions the lowest truth value, namely 0, and tautologies the highest, namely 1. We shall also check whether our theory gives reasonable results in typical cases of scientific inference.

The foregoing is spelt out in an axiom disguised as

DEFINITION 8.1 The structure $\langle S, S_D, [S], \square, \mathord{▯}, \cup, \cap, {}^{-}, \mathscr{V}\rangle$, where S is a nonempty set, S_D a subset of S, $[S]$ the quotient of S by the relation $\Leftrightarrow$ of logical equivalence, $\square$ and $\mathord{▯}$ distinguished elements of $[S]$, $\cup$ and $\cap$ binary operations on $[S]$, $^{-}$ a unary operation on $[S]$, and $\mathscr{V}$ a function on S_D, is called a *metric Boolean algebra of statements* iff

(i) the structure $\langle [S], \Box, \blacksquare, \cup, \cap, ^- \rangle$ is a Boolean algebra, i.e. a complemented and distributive lattice with null element $\Box$ and universal element $\blacksquare$;

(ii) $\mathscr{V}$ is a real valued function on $S_D \subset S$ such that, for any elements p and q of S_D,

$$(a)\ \mathscr{V}(p \,\&\, q) + \mathscr{V}(p \vee q) = \mathscr{V}(p) + \mathscr{V}(q);$$
$$(b)\ \mathscr{V}(p) = 0 \quad \text{for all} \quad p \in \Box;$$
$$(c)\ \mathscr{V}(p) = 1 \quad \text{for all} \quad p \in \blacksquare.$$

Condition (*a*) is common to all metric lattices or measure algebras. Conditions (*b*) and (*c*) determine the range of the values of $\mathscr{V}$. These three conditions are jointly necessary and sufficient to determine $\mathscr{V}$. However, they are not enough to compute the value of an arbitrary propositional compound from the values of its components. (In other words, condition (*a*) is not a full multiplication theorem.) However, we shall see in Sec. 3.4 that this partial indeterminacy is not a serious practical drawback.

Before going any further let us give two warnings. First, contrary to appearances, $\mathscr{V}$ is not a probability measure on S_D, if only because S_D is a set of individuals not a field of sets (a σ-algebra) – as it should if it were to qualify as a probability measure. More on this in Sec. 3.6 point ix. Second, we are dealing with closed bodies of knowledge rather than arbitrary sets of statements – let alone the totality of factual statements. The reason for the restriction is this. Every scientific theory, if consistent, is an ultrafilter (recall Ch. 5, Sec. 3.1). But not all scientific theories in use at a given time are mutually consistent. Hence their union does not constitute an ultrafilter. (Moreover: in general the union of two theories is not a theory.) In other words, Boolean algebra reigns within every theory but does not govern the totality of scientific propositions, not even in a given field of research.

3.3. *Topologies of S_D*

We shall presently show that S_D has two topologies of interest to semantics, given by so many metrics.

DEFINITION 8.2 The function $\delta_- : S_D \times S_D \to [0, 1]$ which assigns to each pair of propositions $p, q \in S_D$ a real number between 0 and 1, such that

$$\delta_-(p, q) = |\mathscr{V}(p) - \mathscr{V}(q)|,$$

is called the *horizontal distance*.

This name for δ_- is not metaphorical, for δ_- has in fact almost all the properties of a distance function, as shown by

THEOREM 8.1 The structure $\mathscr{M}_- = \langle S_D, \delta_- \rangle$ is a quasimetric space, i.e. the distance function δ_- satisfies the following axioms:

(i) $\delta_-(p, q) = \delta_-(q, p)$

(ii) $\delta_-(p, q) + \delta_-(q, r) \geqslant \delta_-(p, r)$

(iii) $\delta_-(p, q) = 0$ iff $\mathscr{V}(p) = \mathscr{V}(q)$

for any p, q, and r in S_D.

The quasi metric δ_- defines a topology in the space S_D. An open ε-neighborhood of $p \in S_D$ is the set

$$U_\varepsilon(p) = \{q \in S_D \mid |\mathscr{V}(p) - \mathscr{V}(q)| < \varepsilon\} \quad \text{with} \quad 0 \leqslant \varepsilon \leqslant 1.$$

This is the set of statements that are equivalent to the given statement to within the tolerance (error) ε. For example, the set of possible confirmers q of a hypothesis p is included in the ε-neighborhood of p. See Figure 8.5.

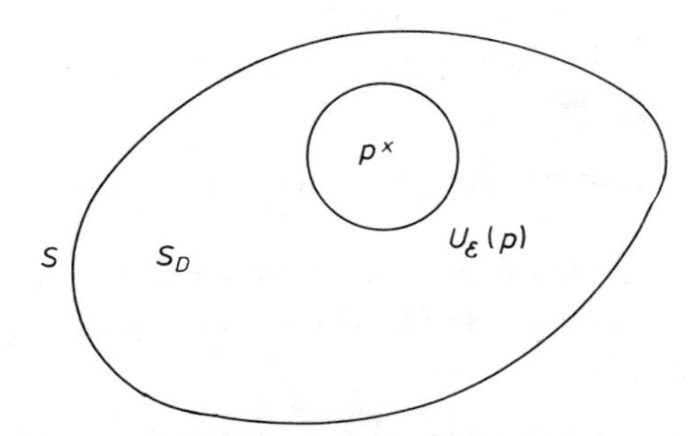

Fig. 8.5. An open neighborhood of $p \in S_D$. All the statements that agree with p to within ε are in $U_\varepsilon(p)$.

We can now formalize the notion of agreement between two statements, which we used in Sec. 2:

DEFINITION 8.3 For any given statement p in S_D and any preassigned real number $0 \leqslant \varepsilon \leqslant 1$, the statement $q \in S_D$ *agrees with* p *to within* ε iff q is in the ε-neighborhood of p.

COROLLARY 8.2 Equivalent statements agree with one another.

Proof. By Theorem 2(iii), the distance between equivalent statements is nil.

Another equally natural topology is determined by a second distance function introduced by the following

DEFINITION 8.4 The function $\delta_| : S_D \times S_D \to [0, 1]$ which to each pair of propositions $p, q \in S_D$ assigns a real number lying between 0 and 1, such that

$$\delta_|(p, q) = |\mathscr{V}(p \vee q) - \mathscr{V}(p \& q)|,$$

is called the *vertical distance*. See Figure 8.6.

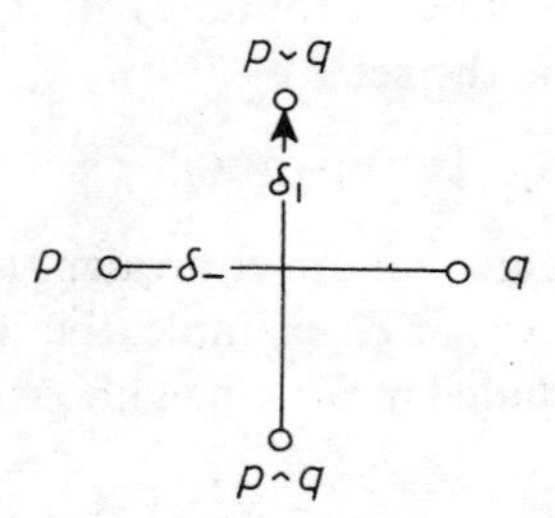

Fig. 8.6. Horizontal and vertical distances between propositions.

This other function deserves its name, as shown by

THEOREM 8.2 The structure $\mathscr{M} = \langle S_D, \delta_| \rangle$ is a quasimetric space.

This new metric defines a second topology in S_D. An open ε-neighborhood of $p \in S_D$ is now

$$U_\varepsilon(p) = \{q \in S_D \mid \delta_|(p, q) < \varepsilon\} \quad \text{with} \quad 0 \leqslant \varepsilon \leqslant 1.$$

The two truth spaces, $\langle S_D, \delta_- \rangle$ and $\langle S_D, \delta_| \rangle$, are separable (Hausdorff) because they are quasimetric. That is, for any two propositions p and q in S_D, there are open sets G and H in S_D such that p is in G and q is in H, and G and H are disjoint. Beyond their shared separability the two truth spaces are quite different, as shown by

THEOREM 8.3 The vertical distance between any two propositions is greater than or equal to their horizontal separation:

$$\text{If} \quad p, q \in S_D, \quad \text{then} \quad \delta_|(p, q) \geqslant \delta_-(p, q).$$

Consequently, any ε-neighborhood built with $\delta_|$ includes the corresponding set constructed with δ_-. Therefore the topology $T_|$ generated by $\delta_|$ is stronger than the topology T_- determined by δ_-. Because the vertical distance between a statement and its negate is maximal (i.e., $\delta_|(p, \neg p) = 1$), which is not the case with the horizontal distance, $T_|$ may be preferred to T_-. This is as far as we go in exploring the topologies for S_D determined by our truth valuation.

3.4. *Comparing Truth Values*

Let us now derive another few consequences from our assumptions. To this end we shall make free use of ordinary logic. And we shall bear in mind that stating a proposition is no indication of its truth value: the latter, if assigned at all, must be assigned by way of an extra metastatement such as, e.g., $\ulcorner \mathscr{V}(p) = \frac{1}{4}. \urcorner$

THEOREM 8.4 For any $p, q \in S_D$, $\mathscr{V}(p \vee q) \geqslant \mathscr{V}(p \,\&\, q)$.

Proof. By Definition 4 and Theorem 2.

THEOREM 8.5 For any $p \in S_D$, $\mathscr{V}(\neg p) = 1 - \mathscr{V}(p)$.

Proof. Set $q = \neg p$ in Definition 1.

THEOREM 8.6 The truth value of the antecedent of a wholly true conditional does not exceed the degree of truth of its consequent:

For any $p, q \in S_D$, if $\mathscr{V}(p \Rightarrow q) = 1$, then $\mathscr{V}(p) \leqslant V(q)$.

Proof. Set $\mathscr{V}(\neg p \vee q) = 1$ in Definition 1 and use Theorem 5 to obtain

$$\mathscr{V}(p) = \mathscr{V}(q) - \mathscr{V}(\neg p \,\&\, q) \leqslant \mathscr{V}(q).$$

COROLLARY 8.3 Equivalent statements in S_D have the same truth value:

For any $p, q \in S_D$, if $\mathscr{V}(p \Leftrightarrow q) = 1$, then $\mathscr{V}(p) = \mathscr{V}(q)$.

Proof. Trading p for q in Theorem 6 yields $\mathscr{V}(q) \leqslant \mathscr{V}(p)$ for the case $\mathscr{V}(q \Rightarrow p) = 1$. This, together with Theorem 6, entails the desired result.

Remark This corollary is not trivial because it is not restricted to formally true biconditionals.

THEOREM 8.7 For any $p, q \in S_D$, if $\mathscr{V}(p \Rightarrow q) = 1$, then

(i) $\mathscr{V}(p \,\&\, q) = \mathscr{V}(p)$
(ii) $\mathscr{V}(p \vee q) = \mathscr{V}(q)$

Proof. By logic,

$$q \Leftrightarrow q \,\&\, (\neg p \vee p) \Leftrightarrow (\neg p \,\&\, q) \vee (p \,\&\, q)$$
$$q \Leftrightarrow q \vee (\neg p \,\&\, p) \Leftrightarrow (\neg p \vee q) \,\&\, (p \vee q).$$

Taking $\neg p \,\&\, q$ and $p \,\&\, q$ as the variables in Definition 1 leads to (i). Similarly for (ii).

In the case of Theorem 7 the truth value of each statement depends on the degree of truth of the other: it is a case of *alethic dependence*. Alethic dependence subsumes *logical* dependence, which obtains when one of the statements entails the other. Let us now tackle the problem of alethic independence. To this end we shall introduce

DEFINITION 8.5 Let $p, q \in S_D$ with $\mathscr{V}(p) \neq 0$. Then the truth value of q relative to p is defined as

$$\mathscr{V}\left(\frac{q}{p}\right) = \frac{\mathscr{V}(p \,\&\, q)}{\mathscr{V}(p)}$$

(Caution: These are not the conditional truth values characterized in Sec. 2.3. In our view all truth values are conditional, i.e., presuppose some base line or other.)

If $\mathscr{V}(p \Rightarrow q) = 1$, Theorem 7 (i) and Definition 5 entail $\mathscr{V}\left(\frac{q}{p}\right) = 1$. In all other cases $\mathscr{V}\left(\frac{q}{p}\right) \neq 1$. Hence $\mathscr{V}\left(\frac{q}{p}\right) - \mathscr{V}(q)$ is a measure of the strength of alethic dependence. This suggests adopting the following

DEFINITION 8.6 Let $p, q \in S_D$ with $\mathscr{V}(q) \neq 0$. Then

(i) p is *alethically independent* of $q =_{df} \mathscr{V}\left(\frac{q}{p}\right) = \mathscr{V}(p)$.
(ii) p is *alethically dependent* upon q iff p is not alethically independent of q.

This relation of alethic independence is not symmetric but, whenever p is independent of q or conversely, the value of their conjunction is the same, namely the product of their degrees of truth. More explicitly, we have

THEOREM 8.8 If p and q are alethically independent statements in S_D, then

$$\mathscr{V}(p \,\&\, q)=\mathscr{V}(p)\cdot\mathscr{V}(q)$$

Proof. By Definitions 5 and 6.

COROLLARY 8.4 If p and q are alethically independent statements in S_D, then

$$\mathscr{V}(p \vee q)=\mathscr{V}(p)+\mathscr{V}(q)-\mathscr{V}(p)\cdot\mathscr{V}(q).$$

We now have all we need in practice: If p and q are alethically dependent we apply Theorem 8, otherwise we use Theorem 9. For purposes of reference we collect these results in the following chart:

p implies q	*p and q are alethically independent*
$\mathscr{V}(p \,\&\, q)=\mathscr{V}(p)$	$\mathscr{V}(p \,\&\, q)=\mathscr{V}(p)\cdot\mathscr{V}(q)$
$\mathscr{V}(p \vee q)=\mathscr{V}(q)$	$\mathscr{V}(p \vee q)=\mathscr{V}(p)+\mathscr{V}(q)-\mathscr{V}(p)\cdot\mathscr{V}(q).$

An obvious application of these results is the following

THEOREM 8.9 Let T be a scientific theory with n independent assumptions A_i. Then

(i) the degree of truth of the axiom basis equals the product of the partial degrees of truth:

$$\mathscr{V}\left(\bigwedge_{i=1}^{n} A_i\right)=\prod_{i=1}^{n}\mathscr{V}(A_i);$$

(ii) the degree of truth of an assumption conjoined with any of its logical consequences equals the former:

$$\text{If} \quad A_i \vdash t \quad \text{then} \quad \mathscr{V}(A_i \,\&\, t)=\mathscr{V}(A_i).$$

Proof. Part (i) follows from an obvious generalization of Theorem 8 to a conjunction of an arbitrary finite number of independent statements. Part (ii) is an application of Theorem 7 (i).

Since a well organized theory is constituted by a bunch of assumptions and all of their consequences, the preceding theorem justifies adopting the following convention concerning the degree of truth of a scientific theory.

DEFINITION 8.7 The *degree of truth of a scientific theory* equals the product of the truth values of its initial assumptions provided the latter are mutually independent.

This definition elucidates the notion of degree of truth of a theory without however allowing us to compute the degree of truth of any nontrivial theory in factual science. Such a numerical value must go uncomputed. All we can do is to estimate the truth value of a few logical consequences of the axioms conjoined with subsidiary assumptions and empirical data, and see whether the results confirm or infirm the premises, both theoretical and extratheoretical. In fact consider the following rather typical process. (For details see Bunge, 1967a, Ch. 15, and Bunge, 1973b, Ch. 10.)

Initial theoretical assumptions: A_1, A_2.
Additional premises: subsidiary hypothesis s and datum e.
Deduce a testable consequence t*:* $A_1, A_2, s, e \vdash t$.
Produce a new empirical datum e' relevant to t.
Contrast t *with* e'.
Estimate the truth value of t assuming e' to be true.
Infer whether the preceding step confirms or infirms the premises.

The attempt to go upstream and calculate the truth value of the initial assumptions on the basis of the degrees of truth of a few of their consequences conjoined with alien extra assumptions (such as s and e) is chimerical.

We close this subsection by defining some related concepts.

DEFINITION 8.8 Let $T = \langle S, \vdash \rangle$ and $T' = \langle S', \vdash \rangle$ be two theories with a common meaning core, i.e. such that $\mathcal{S}(T) \cap \mathcal{S}(T') \neq \emptyset$ and $\mathcal{R}(T) \cap \cap \mathcal{R}(T') \neq \emptyset$. Then T is *truer* than T' iff there is a 1:1 transformation f of S into S' such that, for every p in T, $\mathcal{V}(p) \geqslant \mathcal{V}(f(p))$.

DEFINITION 8.9 Let $T = \langle S, \vdash \rangle$ and $T' = \langle S', \vdash \rangle$ have a common meaning core. Then T and T' are *alethically equivalent* iff there is a truth preserving transformation $f: S \to S'$, i.e. one that, for every p in T, $\mathcal{V}(p) = = \mathcal{V}(f(p))$.

DEFINITION 8.10 Let T and T' be two alethically equivalent theories. Then T and T' are *semantically equivalent* iff they have the same sense and the same referents.

3.5. *Scientific Inference*

Let us now check whether our theory squares with usual procedures of scientific inference. To begin with the theory condones the scientist's common belief that factual propositions are more or less true rather than either wholly true or totally false. Moreover our theory gives quantitative precision to the idea of partial truth. In particular we have the following code.

Scientific vernacular	*Metascientific jargon*
p is *true*	$\mathscr{V}(p) \doteq 1$
p is *approximately true*	$0 \ll \mathscr{V}(p) < 1$
p is *true to within* $\varepsilon > 0$	$\mathscr{V}(p) = 1 - \varepsilon$
p is *partially true*	$\frac{1}{2} < \mathscr{V}(p) < 1$
p is *false to within* $\varepsilon > 0$	$\mathscr{V}(p) = \varepsilon$
p is *nearly false*	$0 < \mathscr{V}(p) \ll 1$
p is *false*	$\mathscr{V}(p) \doteq 0$
p is *truer than* q	$\mathscr{V}(p) > \mathscr{V}(q)$
p and q *agree to within* $\varepsilon > 0$	$\lvert\mathscr{V}(p) - \mathscr{V}(q)\rvert \leqslant \varepsilon$
p and q *disagree to within* $\varepsilon > 0$	$\lvert\mathscr{V}(p) - \mathscr{V}(q)\rvert \geqslant \varepsilon$,

where '$\doteq$' is a standard symbol for approximate equality, and ε the discrepancy introduced in Sec. 2.3.

Secondly, the theory contains the modus ponens and the modus tollens, which are the cornerstones of the theory and practice of deduction. In fact, if $\mathscr{V}(p \Rightarrow q) = 1$ and $\mathscr{V}(p) = 1$, then by Theorem 7(i) $\mathscr{V}(p \,\&\, q) = \mathscr{V}(p)$. Replacing these values in Definition 1 we get $\mathscr{V}(q) = 1$. Similarly for the modus tollens: setting $\mathscr{V}(q) = 0$ in Theorem 7(ii) yields $\mathscr{V}(p \vee q) = 0$, which substituted into Definition 1 entails $\mathscr{V}(p) = 0$. Similar results are obtained by substituting $\doteq$ for $=$, i.e. for the weakened inference patterns.

Suppose now a conditional is asserted tentatively and its consequent is tested for and found to be true (or false) to within a certain discrepancy ε, where $0 < \varepsilon \ll 1$. That is, we have

(i) *Confirmation*

Assumptions: $\mathscr{V}(p \Rightarrow q) = 1,\ \mathscr{V}(q) = 1 - \varepsilon$
Consequence: $\mathscr{V}(p) = 1 - \varepsilon - \mathscr{V}(\neg p \,\&\, q) \leqslant 1 - \varepsilon$

(ii) *Refutation*
Assumptions: $\mathscr{V}(p \Rightarrow q) = 1,\ \mathscr{V}(q) = \varepsilon$
Consequence: $\mathscr{V}(p) = \varepsilon - \mathscr{V}(\neg p \,\&\, q) \leqslant \varepsilon.$

In words: If q is confirmed, then p can be assigned an *upper bound* less than or at best equal to the value of q. And if q is refuted so is p. In short we retrieve what we knew all along, namely that, whereas confirmation is inconclusive, refutation is pretty clear cut – that is, if we restrict our considerations to pairs of statements.

The previous results hold only for *isolated* statements – the normal situation in inductive logic but an extremely artificial one in science. In live science hypotheses are assigned truth values in the light of both further hypotheses and whole bodies of empirical evidence; likewise the latter are estimated in the light of both further evidence and hypotheses –actually whole theories. When this circumstance is taken into account it becomes possible to obtain bounds other than those calculated above: i.e., confirmation may then be strengthened and refutation weakened – or conversely. In short, the whole body of relevant knowledge is summoned to pass judgment on every single hypothesis and every single datum. It is not that *das Wahre ist dans Ganze* (Hegel), but that the *recognition* of truth and of falsity calls for a whole battery of tests (Bunge, 1961c, 1967a, Vol. II, Ch. 15). The whole may be fully meaningful but, if factual, cannot be totally true.

Finally we note that our theory of truth is contiguous with the theory of scientific inference and, in particular, with the calculus of errors of observation. The latter is in charge of assigning the error or discrepancy ε a numerical value.

3.6. *Comments*

(i) The truth measure $\mathscr{V}$ is a continuous function. But this does not allow one to replace the tautology/nontautology dichotomy by a richer gamut of degrees of *logical* truth and falsity. This Aristotelian dichotomy is very basic: it is built into the very algebra of statements, which is a Boolean algebra with just two selected elements, □ and ▯. *This* is what we take "two valued" to mean in reference to ordinary logic – not that truth values other than 0 and 1 are disallowed. In order to introduce (in a purely syntactic fashion) degrees of analyticity intermediate between tautology and contradiction one would have to modify that algebraic structure – e.g., by increasing the number of distinguished elements of the statement set and characterizing them appropriately. However, it is doubtful that such a reform, though algebraically feasible, would be of interest to the logic of factual science. In any case the truth free formulation of logic we have adopted reminds one that logical truth and falsity, unlike factual truth and falsity, are structural or algebraic – something that the model theoretic presentation of logic tends to becloud. Logics are not truth calculi but entailment calculi.

(ii) The above mentioned continuity of the valuation function $\mathscr{V}$ allows one to consider arbitrarily close approximations to either complete truth or total falsity. A typical example is afforded by any series expansion. When expanding a function as a convergent series and adding up only the first n terms, one makes the error $R_n = |S - S_n|$, where S is the exact but perhaps unknown sum of the series while S_n is the known or knowable sum of its first n terms. As more terms are added the remainder R_n decreases and the truth value of the approximation increases accordingly. Indeed, one may set $\mathscr{V}(S_n \mid S) = |1 - R_n/S|$.

(iii) Our theory enables one to elucidate the intuitive notion of the "asymptotic approaching to total truth" – sometimes also expressed, misleadingly enough, as "the gradual advance from relative truth to absolute truth". One way of exactifying this notion is, following Reichenbach (1949), to apply the standard notion of convergence of a sequence to a sequence $\langle p_n \mid n \in N \rangle$ of propositions all with the same form and referent – e.g., the successive outcomes of measurements of the electron charge. This can easily be done but is of no great help because we never have infinite sequences of factual statements of that kind, every one of which has been assigned a truth value. More often than not we switch our interest from one family of propositions to another before we have had a chance to form a sequence long enough to suggest (rather than exhibit) any convergence properties (Bunge, 1963a). Moreover, the corresponding sequence of truth values, i.e. $\langle \mathscr{V}(p_n) \mid n \in N \rangle$, could not possibly exhibit convergence in the strict mathematical sense because the terms of the sequence obey no law. Surely there are sets of theories with the same referent and constituting finite increasing sequences, but these are not very long and there is no reason to believe that any of them could continue indefinitely. In any case there is no confirmed theory of knowledge containing a law statement – not just a Panglossian opinion – to the effect that every sequence of hypotheses (or of theories) concerning any given factual referent should converge to total truth. What we do have is a few generalizations with a restricted scope, such as this one: "The truth value of a statistical estimate approaches unity as the sample size approaches the total population". But this approaching of partial to complete truth is not uniform or lawful, hence it cannot be described with the help of the mathematical concept of limit. Thus if the population consists of *A*'s and *B*'s in the same proportion, sampling may result in picking sheer *A*'s during the first half of the time and only *B*'s the rest of the time, so that during the first period the relative frequency of *A*'s will be 1 instead of the long run frequency $\frac{1}{2}$.

(iv) The valuation function $\mathscr{V}$ is "external" to the algebra of propositions. This tallies with the fact that factual truth values are assigned rather than extracted from the propositions themselves by sheer force of analysis. The latter procedure works only, as noted by Leibniz, for logical truths and falsities. The externality of $\mathscr{V}$ to the algebra has the advantage that we might try functions other than the one determined by Definition 1 while keeping the logic intact. Moreover one might study the whole lot of continuous measures on S_D.

(v) Our claim that truth values, if factual, do not inhere in propositions but are bestowed upon them *ab extrinseco* agrees with our treating truth values, when factual, as values of a certain function $\mathscr{V}$ rather than as elements of the algebra of statements. That factual truth values are assigned (and reassigned) rather than disclosed by logical analysis does not entail that our theory elucidates a pragmatic or methodological concept of truth. It does entail that any application of our alethics calls for methodology. A pragmatic concept of truth would be characterized by (*a*) some function $\mathscr{V}_p$, other than our $\mathscr{V}$, on the set $S_D \times P$ of statement-person pairs, and (*b*) certain assumptions concerning the *P*'s, in particular their truth value assignment habits, or else norms.

(vi) We have defined $\mathscr{V}$ on a proper subset S_D of the totality S of statements. That is, we abstain from assigning truth values to a number of statements – e.g. for lack of relevant evidence or for the absence of a proof. (This aspect of our semantics might be approved of by the mathematical intuitionist.) The reason is simple: as a matter of fact most statements in factual science go unvalued. In short, our alethics makes room for what have been called *truth value gaps* – without requiring any change in logic. It only requires the notion of a partial function, which can be construed as a total function on a domain enriched with a fictitious element embodying the undefined (Scott and Strachey, 1971).

(vii) An alternative solution to the problem of the truth value gaps is of course the adoption of intuitionist logic. But this would require the reconstruction of the whole of mathematics in intuitionist terms, since factual science utilizes, in principle, the totality of mathematics. Our solution to the problem is far less expensive.

(viii) Another solution to the problem of the truth value gaps is to adopt some system of three valued logic involving a third truth value – call it "indeterminate". This move was actually proposed in relation to quantum mechanics, which – like every other theory for that matter – contains empirically untestable statements, such as those referring to the "interphenomena" or events assumed to occur between observations (Reichenbach, 1944). But this proposal, like similar proposals made by a number of authors for different nonreasons, is not workable. First, one should be able to put his finger on at least *one* proof in quantum mechanics requiring rules of inference other than those consecrated by ordinary logic. Second, quantum mechanics should then be reaxiomatized on the basis of the alternative logical calculus. None of these two conditions has been satisfied: this logical revolution is still, after four decades, in the initial proclamation stage. And if it became victorious if would asphyxiate quantum mechanics by isolating it from the rest of physics, which would presumably keep its classical logic. In fact, theories with different logics cannot be conjoined, as they must be if they are to become applicable and testable. (See Bunge, 1973a, and 1973b.)

(ix) The metric Boolean algebra introduced by Definition 1 should not be mistaken for a probability measure. For one thing the arguments of $\mathscr{V}$ are individuals (propositions) not sets. (In other words the domain of $\mathscr{V}$, unlike that of *Pr*, is not a field of sets.) For another, our normalization condition is $\mathscr{V}(p)=1$ for $p \in \Box$, not $\mathscr{V}(S_D)=1$ as it should in order to be a probability measure. Thirdly, our theory contains no semantic assumptions asserting that "$\mathscr{V}(p)$" represents the probability of the proposition p – whatever this expression may mean if indeed it means anything. (It cannot mean "the probability of being true", for this would drag us in a circle: the statement "$Pr(\mathscr{V}(p)=1)=r$" involves the concept of total truth.) This semantic trait of our theory should suffice to distinguish it from the various probability (or improbability) theories of truth even if we were to choose exactly the same mathematical formalism.

There is more: whereas our axioms can be applied right away to situations of interest in real science, the axioms of measure theory yield no probabilistic statements without further ado. Indeed, probability theory proper, as distinct from measure theory, starts where the latter leaves off, namely by specifying (constructing) a probability space or space of "events". (Our basic set $[S]$, on the other hand, is sufficiently characterized by saying that it forms a Boolean algebra.) To put it another way: measure theory provides only the foundations of the probability calculus. These foundations cannot be activated to yield probabilistic results, such as the laws of large numbers, unless enriched with some definite *model* of a possible (though idealized) situation, such as the coin model, an urn model, or a Markov chain model. No such model no probability theory proper. (See, e.g., Kolmogoroff, 1963; Feller, 1968.) And such specifically probabilistic devices are as alien to our theory of truth as they are to measure theory: both truth and measure are independent of randomness. Moreover, all such probabilistic models do is to allow one to assign probabilities to the elementary "events" (the individual sets), never to the corresponding statements. The probability statements themselves, if theoretical, are supposed to be totally true and derived in accordance with ordinary logic. (Their factual truth value is of course another matter.) The widespread view that the theory of probability is a generalization of logic involving probability implications and condoning inductive reasoning ignores the perfectly classical logical structure of the theory. There is somewhat more to all this: we shall look into it in the next section.

(x) Our theory of truth explains what is wrong with Frege's conception of a predicate as a function from individuals in a domain D to truth values (Ch. 1, Sec. 1.3). If we simplify, pretending (with Frege) that every proposition can be assigned a truth value (i.e. if we set $S_D = S$) and keep only the two extremes 0 and 1 of the unit interval, we get the functions

$$P: D \to S \quad \text{and} \quad \mathscr{V}: S \to \{0, 1\}.$$

The composition of these functions yield what may be called the *Frege predicate* F corresponding to the predicate P:

$$F =_{df} \mathscr{V} \circ P: D \to \{0, 1\}.$$

What Frege did was to skip the preceding analysis of F into $\mathscr{V}$ and P.

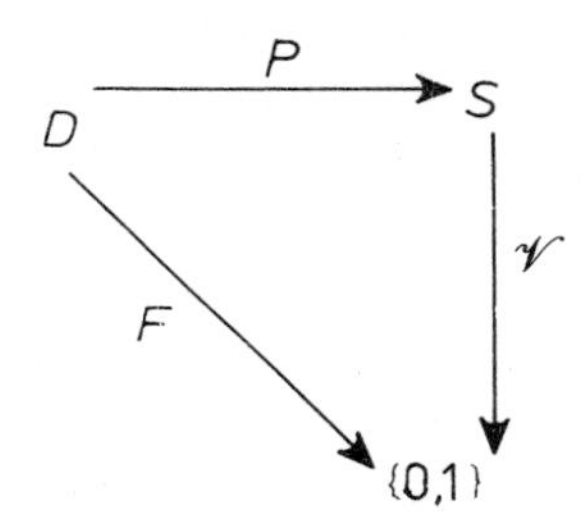

4. Truth et Alia

4.1. *Truth and Probability*

Skeptics and empiricists have rightly insisted on the "probable" nature of factual knowledge. But, until recently, they employed the term 'probable' in a nontechnical acceptation, namely as a synonym for either 'uncertain' or 'corrigible' or both. The probability theories of truth proposed over the past six decades appeared to formalize that view by taking the popular 'probable' in the technical sense that the theory of probability assigns to it. More precisely, these theories equate the degree of truth of a factual statement with its probability or with some increasing (or decreasing) function of the latter. They are thus exact philosophical theories. But they are also hollow, for there exists no procedure, other than arbitrary decree, for assigning probabilities to statements. The passion for exactness is certainly a noble passion but, like any other, it may make fools of us.

Both in applied probability theory and in theoretical science the way to assign primary probabilities is to set up some stochastic model of the *factual* system concerned, such as an urn model. (Recall point (ix) in Sec. 3.6.) This procedure does not work for statements, *e.g.*, scientific hypotheses, if only because these are not picked at random, say by extracting them from a hat full of white (true) and black (false) statements. The very concept of randomness, without which the probability theory finds no application, makes no sense for a unique and carefully contrived object such as a scientific hypothesis.

What can often be assigned probabilities are the *facts* referred to by a (factual) probability statement. Thus we may be able to compute, with the help of a *specific* stochastic theory, or to measure, with the help of a *specific* experimental set up, the probability of a certain *event* belonging to a uniform class of factual items, such as rainfalls in some area. But such probability assignments will be correct or incorrect to some extent that is independent of the objective probability of the given event. For example, the probability of a certain nuclear event *e* may be exceedingly small, say 10^{-24}, while the degree of truth of this probability assignment may be fairly high, say 0.9. That is, we can have $\mathscr{V}[P(e)=10^{-24}]=0.9$. And we can also have the opposite situation, i.e. a high probability value assignment that is nearly false. In short, probabilities of facts and degrees of truth are mutually independent. Hence there is no way of inferring the degree of truth of a statement from the probability of the fact it concerns, nor conversely. Put in another way: There is no supertheory, concerning both a factual domain and a theory about it, that contains law statements relating facts to our knowledge of them. We do not even know whether such lawful connections between facts (random or otherwise) and our knowledge of them could be found.

In conclusion, since the probability (and the information theoretic) theories of truth are not in a position to assign probabilities to statements, we should stop equating the epistemological 'probable' (='uncertain' or 'corrigible') with the semantical 'partially true'. And it would not hurt looking at the way scientists actually estimate the size of the grain of truth of their theories. They may not use the *word* 'truth' (just as they rarely use the term 'cause') because positivism and conventionalism have given it a bad reputation, but they surely use a (presystematic) *concept* of (factual and partial) truth, as is clear from their relentless search for improved representations of facts.

Besides, the concepts of truth are both *more basic* and *more universal* than the concept of probability. In fact we want to be able to say that a given probability assignment (to a fact) is close to (or far from) the truth. And, while all probability statements have presumably some truth value or other, they are only a proper subset of the totality of statements. In short, there is no surrogate for factual truth, and the theory of truth must preside over both stochastic and nonstochastic factual theories. And alethics as well as applied probability presuppose ordinary logic. Practically every theory does.

4.2. *Truth, Meaning and Confirmation*

In Ch. 7, Sec. 2.2 we stipulated that meaning is a property of constructs and that only those expressions in a language that happen to designate constructs are significant. And in this chapter we have agreed that among constructs only statements may, but need not, be assigned a truth value. Figure 8.7 displays these views.

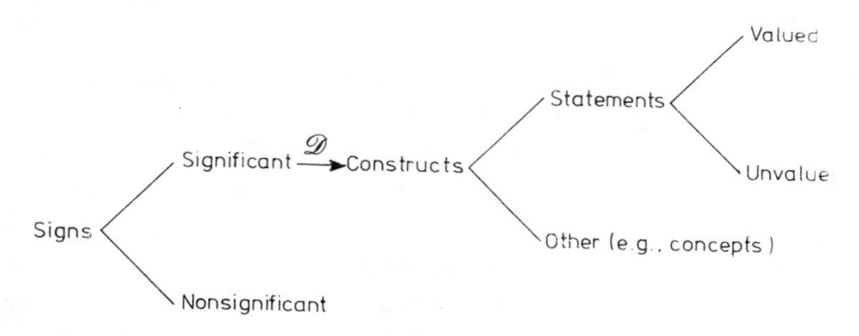

Fig. 8.7. Truth rides on meaning.

Truth depends on meaning (sense together with reference) but not conversely: one and the same factual proposition may be assigned now one truth value, now another, without any change in meaning. (Changes in meaning involve changes in the constructs themselves.) For example, Galilei's refutation of Aristotle's laws of motion did not change the latter's meaning. Galilei could not have accepted either Frege's view that truth values determine meanings or the verifiability doctrine of meaning, according to which meanings are secreted by verification procedures.

However, these doctrines asserting the primacy of truth over meaning survive in various watered down forms. One of these is the view that the truth conditions of a language determine the semantics of the latter. (For the notion of truth condition see Sec. 2.4.) In pragmatic terms: "To give the semantics of a language boils down to stipulating its truth conditions." Let us not quibble over the bizarre belief that languages, rather than theories, come with truth conditions. (For a defense of the alethic neutrality of languages see Ch. 1, Sec. 1.1.) The thesis may fare well in elementary logic but it certainly fails in factual science. First, because in the latter there are hardly any necessary and sufficient conditions under which a statement may be pronounced totally, or even partially, and definitively, or even provisionally, true or false. (Recall Sec. 2.4.) Second, because although every decent scientific theory has a reasonably definite meaning (sense and reference), determined by its basic assumptions, we may be at a loss as to the conditions under which it may be regarded as (approximately) true. In conclusion, the claim that truth conditions determine meanings is as false as the converse thesis. Meaning and truth are equally basic semantic components of propositions.

Nor is truth to be mistaken for confirmation. The two concepts belong in different categories: truth in semantics, confirmation in methodology. But at least in this case taking the one for the other constitutes an intelligent mistake, for confirmation is necessary, though insufficient, to bestow truth values. Indeed, if a statement has been abundantly confirmed by experience and if it coheres with reasonably well corroborated theories, then it may be assigned a truth value close to unity – until new notice. But, as shown by superstition, empirical confirmation alone is insufficient – and refutation may not help. Furthermore empirical tests are sometimes impossible, despite which we may have reasons for allotting a hypothesis a high degree of truth. Figure 8.8 illustrates a frequent situation in theoretical science. (See Bunge, 1967a, I, 5.6.)

What holds for the qualitative concepts of truth and confirmation holds also for their quantitative explicata: confirmation is but an insecure indicator of truth. Thus the statement ⌜All natural numbers are greater than any preassigned natural number⌝ is patently false but, since it has infinitely many confirmers, it could be assigned a maximal degree of confirmation. In sum the two concepts, truth and confirmation, though related, are distinct. A similar relation holds between the semantic con-

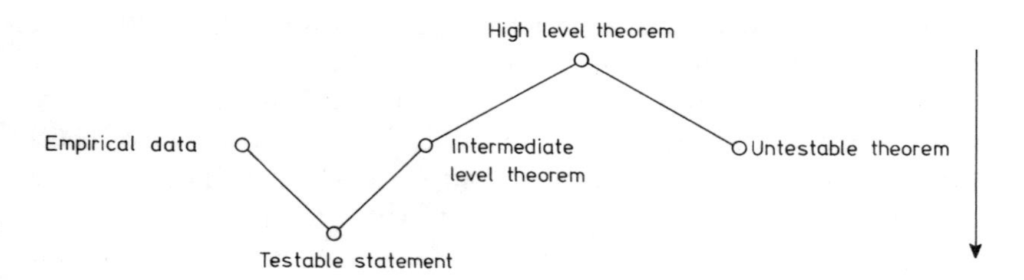

Fig. 8.8. An untestable theorem may be regarded as (more or less) true if it belongs to a theory abounding in corroborated theorems.

cept of truth and the pragmatic (or psychological) concept of acceptance or credence or belief. But this deserves a separate subsection.

4.3. *Truth and Belief*

Except for the dogmatist, truth is not belief: anyone can believe falsities and disbelieve truths – and all of us ignore most truths and most falsities. Let us keep truth and belief distinct – though not necessarily separate. Marry they must if they are to bear fruit or rather edible fruit. The semantic concept of truth and the pragmatic (or psychological) concept of belief (personal or collective) get fused into that of pragmatic truth, or truth for someone. More exactly, there are two such concepts: *personal* (subjective) truth and *collective* (or intersubjective) truth – which, as the fable of the naked king warns us, must not be confused with objective truth. These concepts occur in statements of the form ⌜x believes p⌝, where the substitution instances of x are persons or social groups. The theories dealing with statements of this kind have come to be known as systems of *doxastic logic* and they are closely related to the theories concerning statements of the form ⌜x knows that p⌝, which are the object of the so called *epistemic logic* (Hintikka, 1962).

While there is no doubt that both "logics" are worthy enterprises, it seems clear that they are not *logics* but factual disciplines and consequently in need of empirical tests. Indeed, any belief statement and any knowledge statement of the sort mentioned a while ago concern both propositions and people: they "say" as much or more about persons (or social groups) as about propositions. (They are not metastatements such as ⌜p is true⌝ but object statements with two referents.) This being the case, only psychologists can tell us anything about the truth value of the hypotheses spurned by the doxastic and the epistemic logicians. In other words, although the empirical sciences of belief and knowledge are welcome and moreover long overdue, an *a priori* theory of belief and knowledge stands no chance of being true – hence of contributing to knowledge and compelling our belief in it. To put it another way: Even belief and knowledge statements, i.e., statements of personal (or collective) belief or knowledge, ought to be objective and empirically testable.

Of course objective truths (and falsities) are not Platonic Ideas: *pace* Bolzano there is no such thing as a *Wahrheit an sich*, independent of thinking beings (Bolzano, 1837, I, Sec. 25). Statements in themselves are just useful (as opposed to either idle or evil) fictions. A proposition, viewed from a metaphysical angle, is not an autonomous object (an entity) but perhaps a certain equivalence class of brain processes (thoughts) of some kind. (Note the vagueness, unavoidable at the present stage of cognitive psychology.) And a propositional attitude, such as knowing, believing or doubting a proposition, is another equivalence class of brain processes, this time one concerned with further brain processes. Schematically, we have three levels:

(i) *Class of thoughts* of a kind (judgments).
(ii) *Object proposition* = An equivalence class of thoughts of a kind.
(iii) *Propositional attitude* = An equivalence class of thoughts about thoughts.

(*a*) *p* is *given* – take it or leave it.
(*b*) *p* is *examined* with a view to finding out its degree of truth, or its relation to other propositions, or its usefulness to some action, etc.
(*c*) *p* is *assumed* or hypothesized (not necessarily asserted as true).
(*d*) *p* is *confirmed* (in particular proved) or *infirmed* (in particular refuted).
(*e*) *p* is *assigned a truth value*.
(*f*) *p* is *assigned nonsemantic properties:* systemicity, depth, heuristic power, etc.

(*g*) *p* is *adopted or rejected* for a certain purpose, either theoretical or experimental or practical.
(*h*) *p* is *believed* or *disbelieved* or *left in suspense*.
Etc.

To conclude. Truth and belief are heterogeneous categories. Moreover, *pace* the pragmatists (e.g. Zinov'ev, 1973), truth is not definable in terms of acceptance or credence. Rather on the contrary, to anyone but a mystic, believing *p* consists in admitting *p as true*, at least to a considerable degree at the time of thinking *p*. Which leads us to considering the relation of truth to time.

4.4. *Truth and Time*

If factual truth were an intrinsic property of propositions, as is the case with logical truth, then it would be timeless. And even if it were less than this, namely mere mathematical truth, it would be timeless as well: the property of being true in a model is neither eroded nor accrued in the course of time, because models (in the model theoretic acceptation of the term) are themselves timeless objects. Alas, factual statements are more complicated than this: in addition to being true in some model they must be (sufficiently) true of the world, which is largely beyond our control. And they are assigned truth values on the strength of both conceptual and empirical operations of varying degrees of refinement. The upshot is that factual truth values vary in the course of time. In this sense *Veritas filia temporis* (cf. Bunge, 1967a, II, 10.5.)

The time dependence of factual truth value assignments summarizes a complex process of testing for the truth. This process involves flesh and blood scientists handling a changing array of conceptual and physical tools. Although every one of the inputs participating in this process may be lawful, the output, i.e. the trajectory or curve of the successive truth values does not seem to satisfy any law: it is a time series, at most a trend. (For the concepts of trend and law see Bunge, 1967a, I, 6.2 and 6.6.) This does not mean that truth value assignments and reassignments are capricious, hence that there is no such thing as objective truth. All it means is that, as with most historical processes, sequences of truth value assignments are the outcome of the interplay of numerous factors, some of which escape our control.

What about predictions: do they have a truth value before the facts they refer to come to pass? According to Aristotle (*De interpretatione* Book 9) only propositions about actuals are either true or false. On the other hand propositions about contingent futures have no definite truth values. It might be objected that the predictions computed in factual science do have definite truth values since they are entailed by premises taken to be true. But this would be incorrect: we need not, and usually do not, *assert* our hypotheses but just process them to find out what they entail, and wait and see how some of these consequences fare. That is, we may and often must abstain from assigning any truth values to our predictions. So Aristotle would seem to be right.

However, things are not this simple. If the possibles in question are not unique such as tomorrow's naval battle, but get actualized once and again under suitable conditions, then we may subject them to theory. And we may subsequently contrast some statements in the theory with actuality, or the unfolding of potentiality (to continue Aristotle's train of thought), thus being able to evaluate those propositions. This holds in particular, though not exclusively, for stochastic theories: here we contrast calculated probabilities and their subordinates (averages, mean fluctuations, etc.) with observed frequencies and their relatives. But possibles are not the exclusive property of stochastic theories: all scientific theories concern possible things, possible properties, possible states, possible changes of state. (However, science does not need modal logic. It construes "a possible factual item of kind F" as "an arbitrary member of the set F of factual items", and "a probable fact" as a fact with a definite probability. Hence scientific theory, though concerned with possibility, is strictly truth functional.) Only the applications and tests of scientific theories concern actuals. And no sooner are actuals involved, than we can proceed to an effective (yet provisional) determination of truth values. The foregoing discussion is summarized in Figure 8.9.

5. Closing Remarks

A theory of factual truth is often expected to accomplish at least as much as a theory of formal truth, viz.,

(i) offering a tidy definition of "factually true statement";
(ii) laying down universal truth conditions (criteria); and

(iii) giving rules for computing the truth value of any truth functional compound, such as a conjunction, in terms of the truth values of its constituents.

Our theory of truth fails, nay refuses to undertake the first two tasks and performs only the third. In fact we hold (Sec. 4.2) that the concept

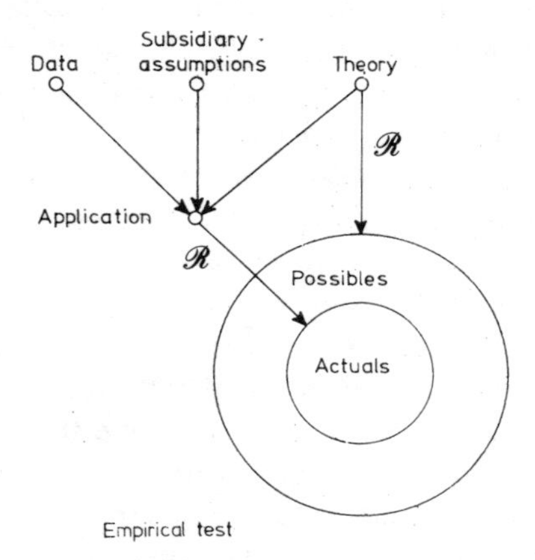

Fig. 8.9. Only the applications and tests of a scientific theory concern actuals: every general theory is about possibles. The outcomes of tests (on actuals) enable one to assign truth values to statements (whether deterministic or stochastic) concerning possibles.

of factual truth is much too basic and universal to be demoted to the rank of a defined concept. (And in any event all the attempts to eliminate it in favor of alternative concepts, such as those of satisfaction, probability, information, and confirmation, have failed.) The most we can do for the concept of factual truth is

(i) to give an *informal* characterization (Sec. 1);
(ii) to show how it is *used* in scientific practice (Sec. 2); and
(iii) to define it *implicitly* by a set of postulates (Sec. 3).

As to factual truth conditions, we have pointed out that (*a*) in science there are no cut and dried truth conditions in the simple form of equivalences, and (*b*) the complex, regional and changing clues for estimating factual truth values must remain the responsibility of science. Here the semanticist must be a spectator and student not a law giver.

Our theory of factual truth, incomplete though it may look to the philosopher accustomed to the white or black situations encountered in logic, does seem to elucidate in an exact way the concept of degree of truth used by applied mathematicians and scientists. However, we do not claim that it is the best possible (the truest) theory. There may well be alternative measures $\mathscr{V}$ that supply a more adequate explication of the intuitive concept of partial factual truth. (Recall Sec. 3.6, point (iv).) Still, any such alternatives should presuppose classical logic, which is the one inherent in science. And they should satisfy the same desideratum, namely that of supplying an exact version of the imprecise notion of approximate truth (or falsity) of fact.

Note finally that our alethics takes care of the notion of *partial explanation*, which is becoming prominent in the philosophy of science (Scheibe, 1973). A fundamental theory is said to provide only a *partial explanation* of the empirical generalizations (or the phenomenological theory) that motivated the construction of the theory. For example, celestial mechanics does not retrieve Kepler's laws but gives only an "approximate explanation" of them because, strictly speaking, the sun does not stand still and the planets exert perturbations upon one another, all of which complicates the actual planet trajectories far beyond Kepler's simple ellipses. As a matter of fact the explanation of the latter supplied by Newton's celestial mechanics is rigorous: what are not exact, or rather not completely true, are the subsidiary assumptions that the solar mass is infinite and that each planet is acted upon only by the sun. Every application of any theory involves simplifying assumptions like these. Thus in the elementary theory of the simple pendulum one assumes that the oscillation amplitude is small and so he obtains the laws of Galilei and Huyghens, known to be only partially true. The deduction is exact: what is approximate is the simplifying assumption and consequently the conjunction of this subsidiary condition with the general assumptions of the theory. To sum up, if we accept the general concept of partial and relative truth of fact we need not introduce the *ad hoc* concept of approximate explanation. Or at least the latter can be defined in terms of the former.

This closes our exposé of the basic theories in our semantics of science. The rest of the book deals with applications (Ch. 9) and border issues (Ch. 10).

CHAPTER 9

OFFSHOOTS

The ideas on meaning and truth presented and discussed in the preceding chapters may be applied to a variety of problems in philosophical semantics, whether pure (general) or applied (to, e.g., scientific theories). We have dealt with sundry problems in applied semantics throughout the foregoing chapters. Let us now face three problems in pure or general semantics, namely those of extension, vagueness, and definite description: in our system all of them presuppose the theories of reference, sense, and truth.

1. Extension

1.1. *The Problem*

An extension is the extension of some predicate. Thus the extension of "– is a mountain" is the class of mountains. And every extension is a set. But not every set is the extension of some predicate. For example there seems to be no (interesting) predicate corresponding to the set {China, d/dx}. Nor does every construct have an extension: only those constructs can be assigned an extension for which the concept of truth makes sense. It makes sense to ask what are the objects for which a certain predicate P holds: the collection of such individuals is the extension of P, or $\mathscr{E}(P)=\{x \mid Px\}$ if P is a unary predicate. Individual concepts, such as "Archimedes", have a referent but not an extension. (Individual variables, such as "here" or "we", fare even worse, as they have no fixed referents.) Nor do sets have an extension – or, if preferred, they are their own extensions. (Sets are in fact the only purely extensional objects.) Not even closed formulas, whether simple such as $\ulcorner Pa \urcorner$, or complex such as $\ulcorner (x)(\exists y)\, Pxy \urcorner$, are usually assigned an extension – although they might. Normally only predicates are assigned an extension or domain of application. At any rate we shall confine the theory of extensions to predicates.

The concept of extension must be elucidated by a theory of extensions. The theory of reference cannot do this job because the concepts of reference and of extension are quite different and probably not interdefinable. For one thing the notion of reference does not presuppose the concept of truth, which the concept of extension does. For another the extension of an n-ary predicate is a set of ordered n-tuples, whereas the reference class of the same relation is the scrambled set of the individuals concerned. Thus

$$\mathscr{E}(\text{loves}) = \{\langle \text{Abélard, Héloïse} \rangle, \langle \text{Dante, Beatrice} \rangle, \ldots$$
$$\mathscr{R}(\text{loves}) = \{\text{Abélard, Héloïse, Dante}, \ldots\}.$$

Prima facie there are two other theories that could claim to constitute a theory of extensions each, namely ordinary logic and set theory. But the former is truth functional rather than "purely extensional", as we argued in Ch. 4, Sec. 1. Logic determines neither intensions nor extensions: it leaves them indeterminate. Only the semantics of logic introduces extensions when supplying models for the logical calculi. (*A fortiori* logic cannot be construed as universal semantics – *pace* Bar-Hillel (1970).)

As to the claim that set theory is the theory of extensions, its force depends upon the version of the theory one has in mind. Certainly it does not hold for the von Neumann-Bernays-Gödel theory, which does not involve the concept of a predicate, whence it cannot be construed as dealing with the extension of predicates: in that theory a class is an object on its own. (The rationale for this is not that "mathematics has no need for non-classes, like cows or molecules" (Mendelson, 1963, p. 160), but rather that, *once generated*, a set may be handled as an object in its own right. But if the problem is to determine or characterize a particular infinite set, then there is no way except to seize on some property and use the principle of abstraction or its rigorous version, the *Aussonderung* axiom – which is absent from the von Neumann-Bernays-Gödel theory.) As to the Zermelo-Skolem-Fraenkel version of set theory, which does involve the notion of a predicate, it is certainly concerned with extensions since it contains the golden bridge between predicates and sets – namely the *Aussonderung* postulate. However, the central goal of this theory is not to elucidate the notion of extension by relating it to that of truth and distinguishing it from that of reference – which is what semantics is mainly interested in.

We need then a separate theory of extensions, distinct from both logic and set theory though subordinated to them. In fact we need two theories of extension if we are to cope with factual science: (*a*) a theory of *strict* extensions, concerning clear cut (decidable) predicates as well as total (not just approximate) truth, and (*b*) a theory of *lax* extensions, concerning predicates that are either inherently vague or whose instances are known imperfectly. The present section is devoted to strict extensions, the next to lax extensions.

1.2. *Strict Extension: Definition*

While the referents of a predicate are the objects it points to, whether truly or not (Ch. 2), the extension of the predicate is the collection of objects for which it holds and in the order in which it holds. More precisely the extension of a predicate P is the *truth set*, or *solution set*, or *graph* of P. This graph is included in the domain of definition of P. *Example 1* Whereas the reference class of "is literate" is mankind, its extension is the subset of persons who actually read and write. *Example 2* Let P be a binary predicate and, more specifically, one whose relata are tied by P in this way: $Pxy=(f(x)=y)$, for x and y in the real line R. The extension of P is the graph of f – a curve in the plane $R \times R$. See Figure 9.1.

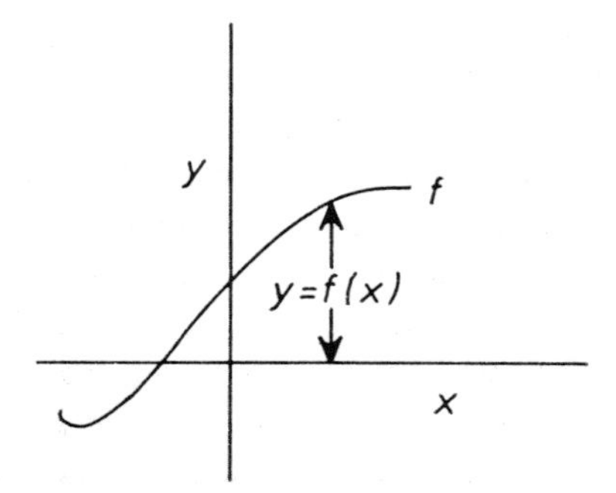

Fig. 9.1. The curve represents the extension of the binary predicate $P=X \times Y \to S$ such that $Pxy=(f(x)=y)$. Indeed, $\mathscr{E}(P)=\{\langle x, y\rangle \in X \times Y \mid y=f(x)\}$.

These ideas are spelled out by the following

DEFINITION 9.1 Let $\mathbb{P}$ be the family of all predicates defined on a domain

$D = A_1 \times A_2 \times \ldots \times A_n$. Then the function

$$\mathscr{E} = \mathbb{P} \to \mathscr{P}(D)$$

such that, for any predicate $P: D \to$Statements, belonging to the predicate family $\mathbb{P}$, $\mathscr{E}$ takes the value

$$\mathscr{E}(P) = \{\langle x_1, x_2, \ldots, x_n \rangle \in A_1 \times A_2 \times \ldots \times A_n \mid Px_1 x_2 \ldots x_n\},$$

is called the *extension function for P* and its values *extensions* (or *extension values*).

Example 1 The extension of "is a planet of the Sun" is the set of the eight planets certified to be such. The ghostly Pluto is among the referents of the predicate and even in its lax extension but it does not qualify, at the time of writing, as a member of its strict extension. *Example 2* A function f and its restriction $f \mid A$ to a set A, though cointensive, have different extensions: $\mathscr{E}(f \mid A) \subset \mathscr{E}(f)$. Obviously this holds, in particular, when A is a singleton.

Remark 1 The extension function $\mathscr{E}$ maps predicates neither into nor onto their domains: the latter contain regions that correspond to different predicates or to none. *Remark 2* Our very notation shows that we reject the extensionalist identification of a predicate with its graph or extension – for the reasons given in Ch. 4 Sec. 1.2. Thus instead of writing '$\langle x, y \rangle \in R$' for a binary relation R, we prefer to write '$\langle x, y \rangle \in \in \mathscr{E}(R)$'. This difference, tiny as it is for practical purposes, is important for the foundations and the philosophy of mathematics as well as for semantics. And it avoids funny expressions such as '$= = \text{diag}\, X \times X$' and '$= \subset \subseteq$'. *Remark 3* The distance between extensions and reference classes can be considerably shortened if, instead of conceiving n-tuples as sets of sets (*à la* Wiener and Kuratowski), we construe them as individuals (points). To this end we may adopt the original Bourbaki point of view, now championed by Mac Lane, according to which an ordered pair is a complex individual characterized by the single axiom (not definition): $\langle x, y \rangle = \langle x', y' \rangle$ iff $x = x'$ and $y = y'$. *Remark 4* Definition 1 bridges Platonism ("Only forms are real") and nominalism ("Only individuals are real"). Or rather it shows that the Platonism/nominalism divide is not a reasonable one, as there are neither pure forms nor formless individuals. Every form is the form of something and every individual exemplifies some form or other. *Remark 5* Because truth is relative, so is extension. The extension of a factual predicate grows or shrinks with the advancement of knowledge.

1.3. *Some Consequences*

First an immediate consequence of Definition 1 in Sec. 1.2:

COROLLARY 9.1 The extension of a predicate is included in its domain:

$$\text{If} \quad P: D \to \text{Statements}, \quad \text{then} \quad \mathscr{E}(P) \subseteq D.$$

Now the main result. But before stating it we must recall that a compound predicate is defined only on the overlap of its components (Ch. 1, Sec. 1.3).

THEOREM 9.1 The extension of a Boolean function of predicates equals the corresponding Boolean function of their images under the extension function. I.e., if P and Q are predicates defined on a common domain, then

(i) $\mathscr{E}(\neg P) = \overline{\mathscr{E}(P)}$;
(ii) $\mathscr{E}(P \,\&\, Q) = \mathscr{E}(P) \cap \mathscr{E}(Q)$;
(iii) $\mathscr{E}(P \vee Q) = \mathscr{E}(P) \cup \mathscr{E}(Q) \supseteq \mathscr{E}(P \,\&\, Q)$;
(iv) $\mathscr{E}(P \Rightarrow Q) = \overline{\mathscr{E}(P)} \cup \mathscr{E}(Q)$.

Proof. For the sake of simplicity let us confine ourselves to unary predicates defined on a domain D. By Definition 1,

$$\mathscr{E}(\neg P) = \{x \mid \neg Px\} = D - \mathscr{E}(P) = \overline{\mathscr{E}(P)},$$

which proves (i). As to (ii), since $P \,\&\, Q$ is defined on a common domain, it can be treated as a single predicate with values $(P \,\&\, Q)\,x$, where $x \in D$:

$$\mathscr{E}(P \,\&\, Q) = \{x \mid (P \,\&\, Q)\,x\} = \{x \mid Px \,\&\, Qx\} = \\ = \{x \mid Px\} \cap \{y \mid Qy\} = \mathscr{E}(P) \cap \mathscr{E}(Q).$$

Analogously for (iii). Finally, (iv) is proved by replacing $P \Rightarrow Q$ by $\neg P \vee Q$ and using (i) and (ii).

Example 1 $\mathscr{E}(\text{nonliving}) = \overline{\text{Living beings}}$.

Example 2 $\mathscr{E}(\text{small circular}) = \mathscr{E}(\text{small}) \cap \mathscr{E}(\text{circular})$.

On the other hand "is a circular square", which refers to plane figures, has a nil extension. (And it is meaningful: recall Ch. 7, Sec. 2.2.)

Remark 1 The condition of Theorem 1, that the component predicates be defined on a common domain, disqualifies the composition (e.g., conjunction) of referentially heterogeneous predicates such as "metallic" and "jealous". Hence we cannot even say of them that they have empty extensions: they have no extensions at all – nor do they have an intension, as we learned in Ch. 4, Sec. 2.2. *Remark 2* Theorem 1(ii) is the basis of the usual procedure for constructing the (extensionally) smallest of all the predicates satisfying a certain condition – namely to conjoin them. For example, consider the class of equivalence relations $\sim_n$, with $n \in N$, on a given set S. The smallest of them all is $\sim_0 = \bigwedge_n \sim_n$. This is the relation that holds only between an element of S and itself – i.e., strict identity.

COROLLARY 9.2 Double negation restores the original extension:

$$\text{If } P \text{ is a predicate, then } \mathscr{E}(\neg\neg P) = \mathscr{E}(P).$$

COROLLARY 9.3 Inconsistent predicates are extensionally void and tautological ones are universal: If P is a predicate with domain D, then

(i) $\mathscr{E}(P \,\&\, \neg P) = \emptyset$;
(ii) $\mathscr{E}(P \vee \neg P) = D$.

Proof. Let $Q = \neg P$ in Theorem 1(ii) and (iii) and use (i). To generalize to arbitrary tautologies and contradictions use the theorem that all tautologies are equivalent, hence coextensive.

Remark 1 All of the above formulas concern predicate variables. Being general they hold for any predicate constant regardless of its reference. For example, if P is assigned the value "philosopher" in Corollary 3, then P refers to persons and is true of philosophers, while $P \,\&\, \neg P$ and $P \vee \neg P$ still refer to persons but now $P \,\&\, \neg P$ "applies" to nothing whereas $P \vee \neg P$ is true of everything. *Remark 2* Corollary 3(ii) and its generalization to an arbitrary tautological predicate is a reason for maintaining that logic is universal strictly speaking, i.e., that it "applies" to everything, or that it holds "in all possible worlds". This could be accepted provided it be not interpreted as the assertion that logic is a sort of universal physics (Gonseth, 1938, p. 20) or metaphysics (Scholz, 1969, pp. 399 ff.). This interpretation is mistaken because (*a*) the referents of the above formulas are predicates (universals) not things, and (*b*) the formulas and rules of logic are consistent with mutually incompatible views of the world. It is the business of metaphysics (or ontology), not of logic, to discover the bricks and plans, if any, of the real world (see Ch. 10, Sec. 4). Logic is not metaphysics, it does not depend on the latter, and it suggests no particular metaphysics (Nagel, 1956). More on this in Ch. 10, Sec. 4.2 and in Bunge (1974a).

An equivalent formulation of Theorem 1 is this: Attributes are isomorphic with their extensions. This is why one can think either in terms of attributes or of their extensions, as well as switch back and forth between them. More precisely, Theorem 1 can be reworded thus:

THEOREM 9.2 Let $\mathbb{P}$ be a family of predicates defined on a common domain D. Then the following holds:

(i) The structure $\mathscr{P}=\langle\mathbb{P}, \square, ▯, \vee, \&, \neg\rangle$, where $\square\in\mathbb{P}$ is the null predicate (the one with null extension) and $▯\in\mathbb{P}$ is the universal predicate (the one that applies to every point in D), is a Boolean algebra.

(ii) The algebra of sets on D, i.e., $\mathscr{D}=\langle\mathscr{P}(D), \emptyset, D, \cup, \cap, {}^{-}\rangle$, is a Boolean algebra.

(iii) $\mathscr{P}$ and $\mathscr{D}$ are isomorphic.

This result had to be stated explicitly because occasionally (e.g., Birkhoff, 1961, p. 189) one reads that there is a dual isomorphism (i.e., an anti-isomorphism) between the algebra of attributes and the one of classes. The duality holds between predicates and their extensions: recall Fig. 4.2 in Ch. 4, Sec. 2.3.

Finally a word on the relation between extension and reference. Except in the case of unary predicates this relation is not simple, because (*a*) extension is reference *cum* truth, and (*b*) whereas reference classes are classes of individuals, extensions are sets of *n*-tuples. A simple relation obtains only between extension and adequate or correct reference. The latter concept is introduced by

DEFINITION 9.2 The *adequate reference* $\mathscr{R}_p^+(P)$ of a predicate P equals the union of the projections of its extension $\mathscr{E}(P)$ on the cartesian factors of the domain of P. In particular, for a binary predicate P defined on $A\times B$,

$$\mathscr{R}_p^+(P)=p_A\mathscr{E}(P)\cup p_B\mathscr{E}(P)\subseteq\mathscr{R}_p(P).$$

1.4. *Comparing Extensions*

Since Definition 1 and Theorem 1 help us to compute individual extensions, we can *a fortiori* compare them. In particular we can ascertain whether two predicates are coextensive in the sense specified by

DEFINITION 9.3 Let P and Q be any predicates. Then P and Q are said to be *extensionally equivalent*, or *coextensive*, just in case they have the same extension:

$$P \sim_e Q =_{df} \mathscr{E}(P) = \mathscr{E}(Q).$$

Remark 1 Identical predicates are coextensive but the converse is false. Thus "lighter than" and "cheaper than" are coextensive in the collection of materials of a given kind but they are intensionally different. If extensionalism were right then coextensives should be identical.
Remark 2 Consequently Leibniz' principle, *"Eadem sunt, quae sibi mutuo substitui possunt, salva veritate"*, is false. In fact let p and p' be two propositions differing only in that the predicate $P' \neq P$ occurs in p' in exactly the same syntactical place P occupies in p. Assume further that P and P' are coextensive. Then p and p' will have the same truth value although different senses: therefore p and p' *non sunt eadem*. Truth values do remain invariant under substitution of coextensives – but truth is not everything and does not determine meaning.

From Definition 3 it is obvious that $\sim_e$ is an equivalence relation. Hence it generates equivalence classes according to

DEFINITION 9.4 Let $\mathbb{P}$ be a family of predicates. Then the class of $\sim_e$-relatives of $P \in \mathbb{P}$ is the subset of $\mathbb{P}$ formed by the equivalence class generated by P, i.e.,

$$[P] = \{Q \in \mathbb{P} \mid Q \sim_e P\}.$$

We can now state a proposition that in a way is the converse of the principle of abstraction or separation of set theory, namely the one predicate-one set axiom. It is the following

PROPOSITION An arbitrary set A determines an equivalence class of predicates under the relation of equiextensionality, namely,

$$\emptyset \subseteq \{Q \in \mathbb{P} \mid \mathscr{E}(Q) = A\} \subseteq \mathbb{P}.$$

Such classes of coextensive predicates need have no structure whatsoever. However the whole lot of such classes does have a definite structure, as will be seen in Sec. 1.5. But we can do better than grouping coextensive predicates: we can compare predicates that are not coextensive provided they are of the same rank. Thus "ellipse" is extensionally included in "conic section", which is in turn extensionally included in "plane figure". In general we have

DEFINITION 9.5 Let P and Q be n-ary predicates. Then P is *extensionally included in* Q iff the graph of P is included in the graph of Q:

$$P \leqslant_e Q =_{df} \mathscr{E}(P) \subseteq \mathscr{E}(Q).$$

We are now in a position to state and prove

THEOREM 9.3 If P and Q are n-ary predicates such that $P \Rightarrow Q$, then P is extensionally included in Q and conversely:

$$P \Rightarrow Q \quad \text{iff} \quad P \leqslant_e Q, \quad \text{i.e.,} \quad \mathscr{E}(P) \subseteq \mathscr{E}(Q).$$

Proof. Assume that Pa holds. Hence $a \in \mathscr{E}(P)$. By hypothesis, if Pa then Qa. Since we did assume Pa, it follows that Qa, which amounts to $a \in \mathscr{E}(Q)$. But a is an arbitrary solution of Px, hence $\mathscr{E}(P) \subseteq \mathscr{E}(Q)$. The converse is proved similarly.

COROLLARY 9.4 Equivalents are coextensive:

$$P \Leftrightarrow Q \text{ iff } P \sim_e Q, \quad \text{i.e.,} \quad \mathscr{E}(P) = \mathscr{E}(Q).$$

THEOREM 9.4 If P entails Q then the extension of P is contained in the extension of Q, i.e., if $P \Rightarrow Q$ is a tautologous predicate then $\mathscr{E}(P) \subseteq \mathscr{E}(Q)$.

Proof. By Theorem 1 (iv), $\mathscr{E}(P \Rightarrow Q) = \overline{\mathscr{E}(P)} \cup \mathscr{E}(Q)$ and, by Definition 1, when $P \Rightarrow Q$ is tautologous $\mathscr{E}(P \Rightarrow Q) = D$. Hence $\overline{\mathscr{E}(P)} \cup \mathscr{E}(Q) = D$, which amounts to $\mathscr{E}(P) \subseteq \mathscr{E}(Q)$.

Let us finally consider the extreme cases of null and maximal extension. More explicitly, we introduce the following two concepts:

DEFINITION 9.6 (i) A predicate with null extension is called *minimal*. (ii) A predicate whose extension coincides with its domain of definition D is said to be *maximal relative to D*.

Example 1 The predicate P such that $Px =_{df} (x^2 = -1)$ & x is a real number, is minimal, for it is satisfied by no real number: $\mathscr{E}(P) = \emptyset$. *Example 2* The predicate P such that $Px =_{df} ((x-1)(x+1) = x^2 - 1)$, is maximal in the field of complex numbers, as it holds for every x in it: $\mathscr{E}(P) = \mathbb{C}$.

THEOREM 9.5 (i) There are infinitely many extensionally minimal predicates of a given rank. (ii) A minimal predicate implies any predicate of the same rank. (iii) There are infinitely many extensionally maximal predicates of a given rank. (iv) A maximal predicate is implied by any predicate of the same rank.

Proof. The first part follows from Corollary 3(i) and the third from Corollary 3(ii). The other two parts follow from Theorem 3 and Definition 6 by recalling that the empty set is included in any set.

In logic and mathematics a minimal predicate is an impossible one: not so in factual science. Here we find plenty of minimal predicates, such as "perfectly rigid", with nonempty reference classes as well as nonvoid senses.

1.5. *Algebraic Matters*

Consider a nonempty set $\mathbb{P}$ of predicates defined on a common domain D. Moreover assume that $\mathbb{P}$ is closed under negation, conjunction, and disjunction. The relation $\sim_e$ of coextension, introduced by Definition 3, effects a partition of the family $\mathbb{P}$ into mutually disjoint equivalence classes $[P]$ characterized by Definition 4. Call

$$\mathbb{P}/\sim_e = \{[P] \mid P \in \mathbb{P}\}$$

the family of all the classes of coextensives in $\mathbb{P}$. We define a partial order $\sqsubseteq$ in this quotient set with the help of the relation $\preccurlyeq_e$ of extensional inclusion introduced by Definition 5:

DEFINITION 9.7 If $P, Q \in \mathbb{P}$, then

$$[P] \sqsubseteq [Q] \quad \text{iff} \quad P \preccurlyeq_e Q.$$

Since $\mathbb{P}$ is partially ordered by $\sqsubseteq$, $\langle \mathbb{P}/\sim_e, \sqsubseteq \rangle$ is a poset itself. Moreover we turn it into a richer structure, namely a lattice, upon defining the lattice operations $\sqcap$ and $\sqcup$ in the following way:

$$[P] \sqcap [Q] =_{df} [P \,\&\, Q], \qquad [P] \sqcup [Q] =_{df} [P \vee Q].$$

In fact consider the class $[P \,\&\, Q]$ of all the predicates coextensive with $P \,\&\, Q$. Since $P \,\&\, Q$ entails P, $[P \,\&\, Q] \sqsubseteq [P]$ by Theorem 3 and Definition 7. Exchanging P for Q we also obtain $[P \,\&\, Q] \sqsubseteq [Q]$. This proves that $[P \,\&\, Q]$ is a lower bound of the subset $\{[P], [Q]\}$ of $\mathbb{P}/\sim_e$. Furthermore $[P \,\&\, Q]$ is the greatest lower bound, or infimum, of that subset. In fact let $[R]$ be a lower bound of the latter, i.e., assume that $[R] \sqsubseteq [P]$ and $[R] \sqsubseteq [Q]$. Then, again by Theorem 3 and Definition 7, $R \Rightarrow P$ and $R \Rightarrow Q$. Now, by logic $R \Rightarrow P \,\&\, Q$, whence $[R] \sqsubseteq [P \,\&\, Q]$. That is, $[P \,\&\, Q] = [P] \sqcap [Q]$ is, indeed, the infimum of $\{[P], [Q]\}$. We proceed analogously with $[P \vee Q]$. This time we start from the logical axiom $\ulcorner P \Rightarrow P \vee Q \urcorner$, to derive $[P] \sqsubseteq [P \vee Q]$ and $[Q] \sqsubseteq [P \vee Q]$. which proves $[P \vee Q]$ to be an upper bound of $\{[P], [Q]\}$. It can also be shown that $[P \vee Q]$ is the least upper bound, i.e., the supremum of the subset of coextensives concerned. Furthermore it can be proved that the equivalence classes in $\mathbb{P}/\sim_e$ inherit the distributivity characteristic of the predicates themselves. Since all of this is proved for arbitrary elements of $\mathbb{P}$ we conclude that the family of classes of coextensives forms a distributive lattice. More precisely, we have

THEOREM 9.6 Let $\mathbb{P}$ be a nonempty set of predicates defined on a common domain and let $\mathbb{P}$ be closed under negation, conjunction and disjunction. Furthermore call $\sim_e$ the relation of equal extension. Then the structure $\langle \mathbb{P}/\sim_e, \sqcap, \sqcup \rangle$ is a distributive lattice.

Moreover this lattice contains a null element $\square$ and a unit element ▯, i.e., equivalence classes such that, for every class $[P]$ of coextensives,

$$[P] \sqcap \square = \square \sqcap [P] = \square, \qquad [P] \sqcup \square = \square \sqcup [P] = [P]$$
$$[P] \sqcap \text{▯} = \text{▯} \sqcap [P] = [P], \qquad [P] \sqcup \text{▯} = \text{▯} \sqcup [P] = \text{▯}.$$

Indeed, since $\mathbb{P}$ is closed under negation and disjunction, we shall find in it at least one tautologous predicate T – one of the infinitely many maximal predicates discussed towards the end of the previous subsection. Then any other predicate S in the family will entail T, i.e., $\vdash S \Rightarrow T$.

Consequently $[S] \sqsubseteq [T]$. This proves $[T]$ to be the maximal element of $\mathbb{P}/\sim_e$: call it $\blacksquare$. Similarly for a contradictory predicate $\neg T$: for every predicate S in $\mathbb{P}$, $\vdash \neg T \Rightarrow S$. Hence $[\neg T] \sqsubseteq [S]$. This proves that $[\neg T]$ is the minimal element of $\mathbb{P}/\sim_e$: call it $\square$. Finally we can build the complement $\overline{[P]} = [\neg P]$ of any class $[P]$ of coextensives: it is such that

$$[P] \sqcap \overline{[P]} = \square, \qquad [P] \sqcup \overline{[P]} = \blacksquare.$$

In short, we have proved that our distributive lattice is complemented and comes with a universal and a null element – in sum, that it is a Boolean algebra. To put it explicitly:

THEOREM 9.7 Let $\mathbb{P}/\sim_e$ be a family of coextensive predicates. Then the structure $\langle \mathbb{P}/\sim_e, \square, \blacksquare, \sqcap, \sqcup, {}^{-} \rangle$ is a Boolean algebra.

This particular Boolean algebra may be called the *Lindenbaum algebra of predicates*, by analogy with the Lindenbaum algebra of propositions. If we disregard all differences between individual predicates and propositions except their differences in extension, we shall limit our attention to their respective Lindenbaum algebras. The extensionalist thesis is, in a nutshell, that only such algebras matter. Our view is, on the other hand, that extensions constitute just one aspect, and not even a basic one, of the concepts of the predicate type. Semantics must investigate all the sides and must show how they are related. We proceed to exhibit the relation between extension and intension.

1.6. *Extension and Intension: The Inverse Law*

The richer a concept the smaller its coverage. Thus since the concept of solid is richer than that of body, the set of solids should be included in the class of bodies. The "inverse law" can be stated in set theoretic terms (Bunge, 1967a, I, p. 68) and can now be proved with the help of the calculus of intensions in Ch. 4 and the concept of extension studied in the previous subsections. It consists in the following

THEOREM 9.8 For any two predicates P and Q of the same rank (hence extensionally comparable)

(i) If $\mathcal{I}(P) = \mathcal{I}(Q)$ then $\mathcal{E}(P) = \mathcal{E}(Q)$;

(ii) If $\mathscr{I}(P) \subset \mathscr{I}(Q)$ then $\mathscr{E}(P) \supseteq \mathscr{E}(Q)$.

Proof of (i). Suppose the consequent of (i) to be false. Then we can "balance" the "inequation" by setting $P = Q \ \& \ R$, where R is a nontautologous predicate such that $\mathscr{E}(P) = \mathscr{E}(Q \ \& \ R)$. By our calculus of intensions (Def. 1(i), in Ch. 4, Sec. 2.2), $\mathscr{I}(P) = \mathscr{I}(Q) \cup \mathscr{I}(R) \supseteq \mathscr{I}(Q)$, contrary to hypothesis.

Proof of (ii). Assume $\mathscr{I}(P) \subset \mathscr{I}(Q)$. Then $\mathscr{I}(Q) = \mathscr{I}(P) \cup X$, with X a nonempty set. Since by hypothesis $\mathscr{I}$ is onto, there exists a third predicate R such that $\mathscr{I}(R) = X$. That is, $\mathscr{I}(Q) = \mathscr{I}(P) \cup \mathscr{I}(R)$. And, again by our calculus of intensions, $\mathscr{I}(Q) = \mathscr{I}(P \ \& \ R)$. Now, the extension of $P \ \& \ R$ is, by the calculus of extensions, $\mathscr{E}(P \ \& \ R) = \mathscr{E}(P) \cap \mathscr{E}(R) \subseteq \mathscr{E}(P)$, i.e., $\mathscr{E}(P) \supseteq \mathscr{E}(P \ \& \ R) = \mathscr{E}(Q)$. In brief, the assumption that $\mathscr{I}(P) \subset \mathscr{I}(Q)$ entails that $\mathscr{E}(P) \supseteq \mathscr{E}(Q)$. Since whatever entails implies, the theorem has been proved.

Remark 1 The converse of Theorem 8(i) is false, as shown by the following counterexample. Let P = Mass density, Q = Specific heat. These two are coextensive, as they apply to all bodies and only to them, but they are not cointensive. *Remark 2* Since the intension of a predicate is included in its full sense, the above theorem holds also for the latter: the richer the sense the poorer the extension. *Remark 3* In our semantics sense and reference, rather than intension and extension, occur on an equal footing: extension depends on reference and truth instead of being a basic characteristic. Moreover extension depends pragmatically, though not semantically, on sense as well. Indeed we cannot proceed to find out the extension of a predicate unless we know its meaning, i.e., its sense and reference. (Try to locate a nondescript object.) *Remark 4* To paraphrase the preceding comment: It is not that extension is a function of sense, but that a *knowledge* of sense precedes an *investigation* of extension. *Remark 5* If there were a semantic relation between intension and extension, other than Theorem 8, we would be able to determine extensions by purely conceptual means: the whole of experimental science would be unnecessary.

Theorem 8 can be given an interesting twist if rewritten in terms of the complements of the extensions, with the help of the theorem: $A \supseteq B$ iff $\bar{A} \subseteq \bar{B}$. Just as $\mathscr{E}(P)$ is the collection of objects for which P holds, so $\overline{\mathscr{E}(P)}$ is the set of objects that fail to satisfy P or, in metaphorical terms, whatever P "excludes". If P happens to have factual referents, then $\overline{\mathscr{E}(P)}$ will be the set of things or facts excluded by P. Reformulated in these terms, our last theorem becomes

COROLLARY 9.5 For any predicates P and Q of the same rank:
(i) If P and Q are cointensive then they exclude the same things;
(ii) The richer and predicate the more it excludes: If $\mathscr{I}(P)\supset\mathscr{I}(Q)$ then $\overline{\mathscr{E}(P)}\supseteq\overline{\mathscr{E}(Q)}$.

If it made sense to assign propositions (not just predicates) an extension, the previous proposition might be regarded as a formulation of Popper's idea that the more it is asserted the more is excluded. But then it would be a trivial restatement of the classical "law".

1.7. *Concluding Remarks*

We have restricted the arguments of the extension function to predicates: in the preceding theory it makes no sense to speak of the extension of a construct of a different category, such as a proposition or a theory. Save exceptions it does not even make sense to ask about the extension of the conjunction of the basic predicates of a theory, since the conjunction of predicates must be defined on a common domain (recall Ch. 1, Sec. 1.3). Thus in a group the group operation, which is binary, and the inverse operation, which is unary, cannot be conjoined to form a third predicate.

However, the restriction to predicates can be lifted in at least two ways. One of them consists in equating the extension of a proposition with that of the most complex predicate occurring in it. For example, the (strict) extension or "domain of validity" of Newton's second law of motion is the collection of bodies with size comprised between that of macromolecules and that of galaxies. Now, the law can be compressed into $\ulcorner(x)\ Nx\urcorner$, where x is the object (or referent) variable and N a complex predicate involving functions and differential operators. Hence we can set

$$\mathscr{E}((x)\ Nx)=\{x \mid Nx\}=\mathscr{E}(N)\,.$$

This extension of the theory of extensions explicates the intuitive notion of the "domain of validity" (or truth range) of a formula, familiar to scientists. The systematic development of this idea is left to the reader.

A second generalization has actually been performed by model theory but it applies only to uninterpreted formulas. The extension of an (abstract) "sentence" s is defined as the collection of models of s, i.e.

$$\mathscr{E}(s)=\{\mathscr{A} \mid \mathscr{A}\in R \;\&\; \mathscr{A}\models s\},$$

where R is the class of all relational structures of a given type. Similarly the extension of an abstract theory, such as Boolean algebra, is the set of all its models. This construal has at least two virtues. One is that it looks natural or intuitive to regard the extension of a formula, or even a whole bunch of formulas, as the totality of its realizations – provided the given formula *has* alternative realizations, i.e., it is abstract. Another is, of course, that if one does this then he gets a ready made theory of extensions, i.e. model theory. When suitably reworded this theory contains our basic Theorem 1 in Sec. 1.3 above (see, e.g., Bell and Slomson 1969 p. 159). But such a general theory of extensions is helpful only with reference to formal constructs and moreover abstract ones. As we saw in Ch. 6, Sec. 2.4, in factual science formulas are already interpreted and satisfied in some fixed mathematical structure or other, so that one is hardly interested in the totality of models of an abstract formula.

We conclude by emphasizing that our theory of extensions is not extensionalist, if only because it is based on a non-Fregean analysis of predicates. The contrast becomes most vivid in the case of a statement such as ⌜All unicorns are stars⌝. From an extensionalist point of view this is a true proposition because it is just an instance of the theorem ⌜The empty set is included in every set⌝. On the other hand in our semantics, which starts from predicates not from their extensions, "not-unicorn or star" does not refer, because its constituents "unicorn" and "star" are defined on disjoint domains. (That this is so constitutes of course a bit of empirical information.) Hence its extension is nil. Therefore the statement is false. And extensionalism does not supply an adequate account of extensions.

2. Vagueness

2.1. *Meaning Vagueness*

Ideally, a scientific predicate should have an exact sense, a precise reference class, and a definite extension. A predicate satisfying the first two conditions shall be called *exact*. If a predicate fails to meet either condition it will be called *inexact*. Note that exactness is not a question of extension for, in our view, truth valuations are external to constructs. Thus we may build a well organized mathematical theory concerning an unheard-of thing, a theory with an exact sense and a precise reference, but which has so far not been assigned a truth value because it has not been subjected to any tests. The peculiar predicates of this theory will then be exact even though they have been assigned no extension.

In the practice of factual science few predicates are exact. Only those belonging to a well organized theory could be exact, but sometimes they are not because of some uncertainty concerning their precise reference. A typical example of referential uncertainty is quantum mechanics, sometimes said to concern individual microsystems, at other times assemblies of such, and most often either individual systems or ensembles manipulated by observers. Under these circumstances quantum mechanical predicates are bound to be inexact in the open context of research even though they satisfy definite mathematical conditions. It is only within a precise formulation of both the mathematical formalism of the theory and its semantics, that the predicates of the theory can be exact. The applied mathematician and the mathematical physicist will not worry about interpretation problems: they will seize on the mathematical formalism shared by all the rival versions. In other words, all the quantum mechanical predicates have a hard core of meaning determined by the formalism (which includes the skeletons of the law statements) and that remains invariant under reinterpretations. This suggests introducing

DEFINITION 9.8 Let P be a predicate shared by every member T of a family $\mathcal{T}$ of theories. Then the *core meaning* of P has the following components:

(*a*) the *core sense* of P:

$$\mathcal{S}_{\text{core}}(P) = \bigcap_{T \in \mathcal{T}} \mathcal{S}_T(P);$$

(b) the *core reference class of* P:

$$\mathcal{R}_{\text{core}}(P) = \bigcap_{T \in \mathcal{T}} \mathcal{R}_T(P).$$

Our definition applies not only to the alternative interpretations of a given mathematical formalism but also to any theories that share a given predicate. For example, while the full meaning of "temperature" is determined by the totality of theories in which it occurs, their intersection determines the core meaning of that predicate. What remains outside the core is precisely the meaning vagueness of the predicate. More precisely, we adopt

DEFINITION 9.9 Let P be a theoretical predicate with given core sense $S_{\text{core}}(P)$ and given core reference class $\mathscr{R}_{\text{core}}(P)$. Then the *vagueness in the meaning* of P relative to the theory T is

$$\Delta_T\mathscr{M}(P)=\langle\Delta_T\mathscr{S}(P),\ \Delta_T\mathscr{R}(P)\rangle$$

where

$$\Delta_T\mathscr{S}(P)=\mathscr{S}_T(P)\ \Delta\mathscr{S}_{\text{core}}(P)$$
$$\Delta_T\mathscr{R}(P)=\mathscr{R}_T(P)\ \Delta\mathscr{R}_{\text{core}}(P)$$

and 'Δ' stands for symmetric (Boolean) difference.

The preceding considerations apply to theoretical predicates only. In such cases the concept of meaning vagueness turns out to be an exact concept. This is not the case with nontheoretical predicates such as "ugly". In this case one might feel tempted to try a topological approach. For example, one might wish to characterize as vague (exact) any predicate that is a limit (inner) point of a given set of predicates. But the very notion of neighborhood, necessary to define limit points and inner points, presupposes the existence of a topology of predicates. And this is forthcoming just in case the predicate family is structured, which was the case studied in Ch. 4, Sec. 2.4, but is not the case with ordinary knowledge predicates. In the latter case meaning vagueness is vague itself.

What should be done about inexact predicates? Either of two things: dressing them up in an exact garb or schooling them until they become exact. The former policy is partially implemented by allowing inexact predicates to adjust to a permissive logic of their own – e.g., some system of three valued logic (Körner 1964). We do not counsel taking this course: meaning vagueness can originate either in woolly thinking or in genuine theoretical discrepancies, and in either case it should be exhibited and worked on rather than shoved under some respectable rug. If we relax the logical standards we shall be unable to exactify our concepts within a theory as well as to discuss their differences when placed in different theories. What ought to be done is to minimize the meaning vagueness within every theory. To achieve this goal there is but one means: to improve on the logical organization and the semantics of our scientific theories. If necessary we should axiomatize them. Of course this will not guarantee the disappearance of vagueness, because alternative axiomatizations are always possible and some of them may not consist in a mere reshuffling of a fixed set of constructs. In other words, a residual meaning vagueness may be unavoidable – not as an indicator of conceptual fuzziness but rather as a healthy token of theoretical variety. While we should wish to maximize *intra*theoretical exactness we should not minimize *inter*theoretical vagueness, for this is achieved by simply outlawing all rival theories but one.

2.2. *Extensional Vagueness*

An inexact predicate is bound to be assigned an imprecise extension. For, if we are uncertain about its meaning, we shall find that there are border line cases. In this case we speak of *extensional vagueness*. A remedy for it is of course exactification. (Always go to the root of the trouble.) For example, by substituting a quantitative concept of length for the qualitative concepts "long", "medium" and "short", we get rid of meaning vagueness and at the same time decrease extensional vagueness. However, the latter may not be shrunk to nought except in very simple cases, because in general we shall have shades of truth values instead of clear cut cases of truth and falsity. Consequently extensional vagueness is sufficient but not necessary for inexactness: the former may originate in the uncertainty inherent in our truth value assignments. For this reason it is convenient to introduce a concept of extensional vagueness independent of the one of meaning vagueness elucidated in the preceding subsection.

The *strict* extension of an exact predicate P with domain D is the class of objects in D for which P is true:

$$\mathscr{E}(P)=\{x\in D \mid Px\}, \quad \text{or} \quad \mathscr{E}(P)=\{x\in D \mid \mathscr{V}(Px)=1\}.$$

The generalization to partial truth gives rise to the notion of lax extension:

DEFINITION 9.10 Let P be a predicate with domain D and $\mathscr{V}$ a truth valuation, while ε is a preassigned real number comprised between 0 and 1. Then the *extension of P to within* ε is defined as

$$\mathscr{E}_\varepsilon(P)=\{x\in D \mid 1-\varepsilon\leqslant\mathscr{V}(Px)\leqslant 1\}.$$

This concept of lax extension covers the two cases we discussed above: that of extensional vagueness due to inherent meaning vagueness, and the one due to uncertainties in truth valuation. Lax extensions include strict extensions:

$$\text{For every } 0\leqslant\varepsilon\leqslant 1,\ \mathscr{E}_\varepsilon(P)\supseteq\mathscr{E}(P).$$

The excess of the former over the latter is precisely the amount of extensional vagueness. More explicitly, we propose

DEFINITION 9.11 Let P be a predicate with domain D, lax extension $\mathscr{E}_\varepsilon(P)$, and strict extension $\mathscr{E}(P)$. Then the *extensional vagueness fringe* of P is

$$\Delta\mathscr{E}(P)=\mathscr{E}_\varepsilon(P)-\mathscr{E}(P)=\{x\in D \mid 1-\varepsilon\leqslant\mathscr{V}(Px)<1\}.$$

In words: $\Delta\mathscr{E}(P)$ includes all and only the border line cases.

The same ideas can be restated in terms of a generalized membership relation $\in_v$ defined as follows:

$$x\in_v S=_{df}\mathscr{V}(x\in S)=v, \quad \text{with} \quad 0<v\leqslant 1.$$

It is easily seen that, if $x\in_v S$, then $\neg(x\in_v S)=x\in_{1-v}S$. In terms of this generalized membership concept, the extensional vagueness fringe of a predicate P with domain D and strict extension $\mathscr{E}(P)$ becomes

$$\Delta\mathscr{E}(P)=\{x\in D \mid x\in_v\mathscr{E}(P) \quad \text{and} \quad 0<v<1\}.$$

In other words, our very sets may be allowed to be fuzzy to some extent by tying the notion of membership to that of partial truth. (For a different approach see Goguen (1969).)

Still another approach consists in focusing on statements rather than their referents. Consider an obviously vague predicate such as "healthy", or H for short, defined on the set O of organisms, i.e., $H:O\rightarrow$ Statements.

Then H induces a tripartition of this set S of propositions according as the propositions Hx, with $x \in O$, are true, false, or neither. The extensional vagueness fringe of H may then be defined as the set of statements of the form Hx which prove to be neither true nor false – i.e., the statements concerning all the border line cases. If this view is adopted then the amount of extensional vagueness is given by the fraction of such alethically indeterminate statements. (Caution: do not construe such a fraction as a probability. Statements are not random variables and truth values are not allotted at random.)

We shall not pursue this matter any further but shall instead refer to the recent literature for alternative approaches (Körner, 1964; Bunge, 1967a; Gentilhomme, 1968; Goguen, 1969; Castonguay, 1972; Moisil, 1972). All of them share the conviction, laughed away by inexact philosophers, that "vagueness" can be exactified even if vagueness cannot be shrunk. There is no more paradox in this view than in the mathematical theory of approximations – nor more than in the thesis that "exact" can always be rendered more exact.

2.3. *Structural Indefiniteness*

There is a deeper rooted kind of imprecision for which no easy remedy is in sight, namely what we shall call *structural indefiniteness.* First the intuitive notion. A negative statement is less definite or less committed than an affirmative one, an existential statement less so than a universal generalization, and a possibility statement far less definite than the corresponding nonmodal proposition. Unlike the types of vagueness we have investigated before, structural indefiniteness is not due to predicate inexactness or to uncertainty in extension, but seems inherent in logical form.

A possible way of assigning indefiniteness values and thus elucidating the notion of structural indefiniteness is by adopting the following principles.

1 Atomic statements have zero structural indefiniteness.

2 The degree of structural indefiniteness of a molecular statement p equals the number of negations plus the number of disjunctions occurring in p.

3 Possibility statements, though indefinite, have no definite structural indefiniteness value.

Accordingly, for affirmative atomic statements p and q, we have

$$Ind(p)=0, \qquad Ind(\neg p)=1$$
$$Ind(p \,\&\, q)=0, \qquad Ind(p \vee q)=1$$
$$Ind(p \Rightarrow q)=2, \qquad Ind(p \Leftrightarrow q)=4$$
$$Ind\left(\bigwedge_{i=1}^{n} Fx_i\right)=0, \qquad Ind\left(\bigvee_{i=1}^{n} Fx_i\right)=n-1.$$

This concept of indefiniteness or weakness of commitment is of importance in the methodology of science (see Bunge, 1967a, I, pp. 273ff). It is often conflated with those of logical strength, content, and improbability (see Ch. 4, Sec. 3.2). Which is a reflection upon the state of semantics.

3. Definite description*

3.1. *The Received View: Criticism*

Definite descriptions, such as 'my mother', 'the logarithm of 1', and 'the saint next door', can pose some subtle logical and semantical problems – particularly when they refer to something that happens not to exist. It took *the* Russell to realize this and to try and analyze definite descriptions with the help of the then young mathematical logic. Russell's now classical "theory", or rather definition, boils down to this: A definite description presupposes existence and indicates uniqueness (Russell, 1905, 1919a). That is, "the so-and-so is such-and-such" must be analyzed as "There is a unique thing that is so-and-so and such-and-such". In symbols,

$$G((\imath x)\, Fx) =_{df} (\exists x)\,(Fx \,\&\, Gx \,\&\, (y)\,(Fy \Rightarrow y=x)) \qquad (R)$$

Hilbert and Bernays (1968) gave a different version of the idea that "the so-and-so" is no more and no less than "there is a so-and-so and it is unique". They posited the new rule of inference

$$\frac{\begin{array}{c}(\exists x)\, Fx \\ (x)\,(y)\,(Fx \,\&\, Fy \Rightarrow x=y)\end{array}}{F((\imath x)\, Fx)} \qquad (HB)$$

* Adapted from Bunge (1971c).

subject to certain restrictions. The differences between *HB* and *R* are these: (*a*) while *R* is a definition, *HB* is a rule of inference that, if accepted, must be added to the rules of the predicate calculus; (*b*) *R* is contextual, in the sense that the referent must be assigned a further property (i.e., *G*) to the one it exemplifies uniquely (i.e., *F*); *HB* makes no such demand but as a compensation it forces us to employ correct yet redundant expressions such as 'my mother is a mother' and 'the square of 2 is a square of 2'; (*c*) while in *R* existence and uniqueness are fused together, in *HB* they are stated separately.

Each of the above explications, *R* and *HB*, has its merits and demerits. In particular, *R* is simpler than *HB*, while *HB* involves a neat separation of existence from uniqueness. However, this detachment of the two concepts should be carried even further: one should be able to describe nonexisting objects as well as objects the existence of which has not yet been established. In other words, "it is quite natural to employ descriptions *before* they have been proved to be proper" (Scott, 1967, pp. 181–182). Thus the bachelor should be allowed to say 'My wife does not exist' and the cosmologist should be able to wonder whether there is such a thing as the center of the world. Yet neither *R* nor *HB* render this possible, as both make existence a condition for definite description. For this reason both *R* and *HB* are unsuitable. The same objection holds for some other explications of the definite description (e.g., Kalish and Montague, 1957; Eberle, 1969) – certainly not for Hintikka's (1969).

In mathematics as well as in factual science questions of uniqueness are separate from questions of existence: an object satisfying a certain description may not exist or, if it does, it may not be unique. For example, the theory of differential equations contains separate existence theorems and uniqueness theorems. And in theoretical physics one can often give an unambiguous characterization of a certain object whose actual existence is far from certain: thus one may speculate on the ground state of a mesic atom that has not yet been and may never be brought into existence.

Whether in mathematics or in factual science, when attempting to validate uniqueness claims one proceeds in this way. One first assumes or exhibits existence and then one goes on to assume or investigate uniqueness under the assumption of existence: it would be wasteful to look for unique nonexistents. In other words uniqueness theorems take the form: "If there is an x with the property F then there is no y, other than x, that exemplifies F". Thus uniqueness *proofs* depend on existence assumptions or proofs but not conversely. Similarly for the empirical validation of hypotheses of factual uniqueness. But this does not mean that the *concept* of uniqueness depends upon that of existence – unless of course one is an intuitionist or an operationist. In fact the two concepts are not interdefinable. (If they were then every uniqueness theorem would be just a rewrite of some existence theorem.) And uniqueness assumptions are not deducible from existence assumptions alone.

We submit that *definite descriptions indicate only uniqueness*: that they are by themselves uncommitted as to existence – even though establishing the latter is necessary for proving the former. Otherwise the bachelor could not make jokes about his wife, the atheist could not argue about the Christian god, the physicist could not speculate about the element number 110, and the cosmologist could make no hypotheses about the center of the universe. This being so, we cannot accept any explication of definite description that involves existence. In this respect then *HB* is as inadequate as *R*. We must therefore look for a different characterization. We shall come up with two definitions, neither of which will force us to augment the set of rules of inference.

3.2. *An Elementary Analysis of Definite Descriptions*

The standard analyses of the definite description equates it with existence and uniqueness. Ours will drop existence and retain uniqueness. Now, the uniqueness condition can be expressed in a number of languages. In the present subsection we shall propose an analysis of definite description within first order predicate logic.

We are concerned with uniqueness irrespective of existence, which should be assumed or denied separately. Furthermore we wish to analyze the notion of *relative uniqueness* or uniqueness in some respect, doing it at first with the limited resources of predicate logic. There are two ways of doing this, depending on the "respect" in which an object is unique. An object may be unique because it is the sole instance of a given property, as is the case with the third power of 2. Or the object may be unique because of a relation it bears to some other object, as is the case with my mother. We shall elucidate these two notions by means of the following conventions.

DEFINITION 9.12 The object a is unique in the respect denoted by the predicate $F =_{df}$ a exemplifies F and there are no further individuals, other than a, that exemplify F:

$$a \text{ is } F\text{-unique} =_{df} Fa \,\&\, \neg(\exists x)(x \neq a \,\&\, Fx)$$

or, equivalently,

$$a \text{ is } F\text{-unique} =_{df} Fa \,\&\, (x)\,(Fx \Rightarrow x = a).$$

Next we stipulate that "a is F-unique" amounts to "the x such that x is an F". More explicitly, we lay down

DEFINITION 9.13 The object a is (equal to) the x such that x is an F, just in case a is unique:

$$(a = (\lrcorner\, x)\, Fx) =_{df} a \text{ is } F\text{-unique}.$$

We have chosen '$\lrcorner$' to designate the definite descriptor both for typographic convenience and to forestall confusion with Russell's symbol, which designates a different concept.

The above definitions entail our *elementary explication of the definite description*:

$$(a = (\lrcorner\, x)\, Fx) =_{df} Fa \,\&\, (x)\,(Fx \Rightarrow x = a) \qquad (N)$$

For example, ⌜2 is the smallest prime⌝, i.e., $\ulcorner 2 = (\lrcorner\, x)\, SPx \urcorner$, is now analyzed as: $\ulcorner SP2 \,\&\, (x)\,(SPx \Rightarrow x = 2) \urcorner$, where in turn SPx is defined as $(y)\,(Px \,\&\, Py \,\&\, x \neq y \Rightarrow x < y)$.

For N to hold Fa need not be true. And if Fa is not asserted (separately) then $(\exists x)\, Fx$ does not follow. Consequently N does not involve (entail) existence. For example, the equality

Zeus = The boss of the Greek Olympus.

which is a substitution instance of the left hand side of N, does not commit us to paganism. It is just a designation convention. If challenged, then the right hand side will be challenged as well but the definitory equality will still hold. Likewise if N is construed as an equivalence, since for $\ulcorner A \Leftrightarrow B \urcorner$ to hold both sides must have the same truth value, e.g., falsity. If we now assert (separately) that Zeus is a boss of the Greek Olympus then we conclude that Zeus exists; and if we deny the same statement, i.e., negate Fa, then we commit ourselves to the statement that Zeus does not exist.

So much for definite descriptions in terms of unary predicates. We now generalize our analysis to predicates of any rank. But to keep the exposition readable we limit our definitions to binary relations.

DEFINITION 9.14 The object a is unique in its relation R to $x =_{df}$ a bears the relation R to x and there are no other individuals y, except a itself, that are R-related to x:

$$a \text{ is } R\text{-unique in its relation to } x =_{df} Rax \,\&(y)\,(Ryx \Rightarrow y = a).$$

DEFINITION 9.15 The object a is (equal to) the x such that x bears the relation R to $b =_{df}$ a is unique in its R relation to b:

$$(a = (\lrcorner\, x)\, Rxb) =_{df} a \text{ is } R\text{-unique in its relation to } b$$
$$=_{df} Rab \,\&(x)\,(Rxb \Rightarrow x = a).$$

To sum up, we have identified definite description with uniqueness. Unlike the accepted view ours does not involve the assumption that the described individual exists in some context: it only asserts the *non*-existence of *other* individuals satisfying the same conditions. This is as it should be, for existence is not a matter of definition, let alone of designation: existence is a matter of either assumption or validation. In other words an existence claim is a hypothesis that must be substantiated not a convention that can be freely introduced. For example, whether or not a function with given properties exists, is a matter for an existence theorem to decide. And whether or not a concrete object with assumed properties exists, is a matter that cannot be decided without performing empirical tests. These methodological requirements are violated by the construals R of Russell and HB of Hilbert-Bernays (see Sec. 3.1) but not by ours (N). Indeed, according to our construal the existence of the described individual may, but need not, be asserted separately, namely thus: $(\exists x)$ (x is F-unique), or thus: $(\exists x)$ (x is unique in its R relation to b). Consequently there will be no *logical* differences between a proper (or full) description and an improper (or empty) description: the differences are solely semantic. The semantic peculiarities of the definite description will be taken up in Secs. 3.5 and 3.6 in the light of a deeper yet simpler analysis of our subject, to which we turn presently.

3.3. *A Mathematical Analysis of Definite Descriptions*

We shall now avail ourselves of the general concept of a function, which goes beyond the predicate calculus. Consider the formula ⌜The cost of x is (equals) y⌝ or, more briefly, ⌜x costs y⌝. For every substitution instance of x there is exactly one value of y such that y equals the cost of x. Hence cost is a function – call it C – so that we can write: ⌜$C(x)=y$⌝. Likewise ⌜The father of x is y⌝, or shorter ⌜y fathered x⌝, can be symbolized as: ⌜$F(x)=y$⌝, where 'F' stands for the fatherhood function. These symbols convey the idea that cost and fatherhood are properties of something and moreover that these properties are adequately represented by functions in the mathematical not in the logical sense, for their values are not statements but further individuals. These examples may also be construed as relational statements with the qualification that the relations C and F be many-one. But this construal, suitable for ordinary knowledge cases such as ⌜Scott wrote *Waverley*⌝, is inadequate for most scientific purposes. In science one prefers functional statements of the form "$F(x)=y$".

Let us now *truncate* the functional formula ⌜The F of x equals y⌝ by leaving out the value of the function. We thus get "The F of x", or "$F(x)$", which may be called a *functional semistatement*. This expression indicates the function of interest and an arbitrary value of its argument but not the corresponding value of the function. If the function does have a value at x, i.e., if F is "defined" at x, then that value is unique by definition of the concept of a function. And this is all a definite description indicates, namely a unique object. We compress the preceding into

DEFINITION 9.16 Let F be a function from a set A into a set B. Then the expression '$F(a)$' is called a *proper definite description* $=_{df}$ F is defined at $a \in A$.

If $F(a)$ is a proper definite description then it names a unique individual, say b, in the codomain B of F. That is, the relation between a proper name b and a definite description $F(a)$ is now the full fledged statement "$b=F(a)$", read 'b is the F of a'. In this case $F(a)$ is a name and it poses no further problems. Otherwise we have an improper or empty definite description. More explicitly we lay down

DEFINITION 9.17 The functional semistatement $F(a)$ is an *improper definite description* $=_{df}$ F is not defined at $a \in A$.

Examples: "The weight of my thoughts", "The father of the universe". Suggestion: a number of metaphors are just improper descriptions. This remark might be useful in analyzing the structure of some metaphors.

An alternative but essentially equivalent explication is obtained with the help of the concept of a partial function, or a correspondence between a *sub*set of A and a set B. Thus "king" and "president" are partial functions on the set of countries: "king" is a total function on the set of monarchies and "president" a total function on the set of republics. In general we have

DEFINITION 9.18 Let F be a partial function with domain A. Then $F(a)$ is a *proper (improper) definite description* $=_{df}$ a belongs (does not belong) to A.

On either of the two functional construals, the description can be stated and analyzed without even having to introduce a special symbol. The link between the preceding elementary elucidation (Sec. 2) and the present one is supplied by

DEFINITION 9.19 Let P be a unary predicate with domain B and let F be a function from a set A into B. Then for all $b \in B$

$$b = (\lrcorner x)\, Px =_{df} F(a) = b.$$

Given a unary predicate P it is always possible to find the corresponding function F that will satisfy the preceding convention and conversely. For instance "golden mountain" may be construed not only as a (molecular) predicate but also as a function on the set of places. Consequently assuming that the name of the golden mountain is 'Shiny', Definition 19 gives

Shiny = The golden mountain $=_{df}$ The golden mountain at place a = Shiny.

There is more to the relation between predicates and functions: let us look into it.

3.4. *Continuation of the Analysis*

Take "talented" as an example of a unary predicate P and "author of" as an instance of a function F. (We brush aside the case of joint authorship.) Both P and F are mappings: P maps writers into statements while F maps books into writers. That is,

P Talented: Writers → Statements
F Author of: Books → Writers

In the first case we have, e.g., ⌜Walter Scott is talented⌝ while in the second we may have ⌜The author of *Waverley* is (equal to) Walter Scott⌝. (The *is* in the first statement is predicative while the *is* in the second statement is that of equality.) This allows us to write

$$A(w)=s,$$

where '$A(w)$' stands for "the author of *Waverley*" and 's' for Walter Scott. Consequently ⌜Walter Scott is talented⌝ can be transformed into ⌜The author of *Waverley* is talented⌝, i.e., briefly $T(A(w))$. Now, this is just an instance of the composition of the functions A and T, i.e.,

$$\text{Books} \xrightarrow{\text{Author of}} \text{Writers} \xrightarrow{\text{Talented}} \text{Statements}.$$

This composition may be represented as a commutative diagram.

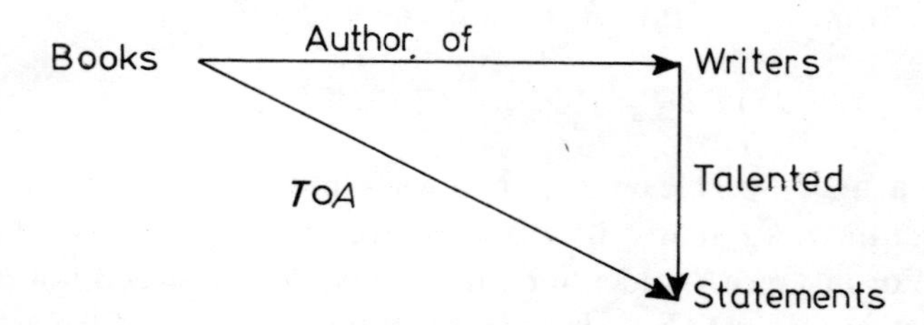

Obviously this works only when the functions concerned are defined at all the points of their domains and when the range of the first equals the domain of the second. This remark will help us to solve a problem concerning the relation between definite descriptions and names.

As Russell pointed out, definite descriptions are not names. They may have the same referent but they do not mean the same. Thus 'The mother of God' "says" more than 'Mary' in the context of Catholic theology. Whatever significance a name may be assigned (and this is a controversial point), the associated definite description, if any, must have a sense containing the previous one. Nevertheless for the purposes of deduction proper definite descriptions can be treated just like names. For example, 'the sum of 2 and 3' can be replaced by the numeral '5'. In other words, although a definite description *is* not a name it behaves in inference *as if* it were a name – provided the necessary precaution is taken. The condition for this functional or behavioral identity of names and definite descriptions is, of course, that the function F in question be defined at the point of interest. For only in this case will $F(a)$ constitute a proper or nonempty definite description according to Definition 16. In other words, a definite description can be treated as a name provided there *exists* an individual complying with the description. Example:

$$\begin{array}{rl} Ts & 1 \\ A(w)=s & 2 \\ \therefore T(A(w)) & 1, 2, \text{Principle of identity.} \end{array}$$

This inference is valid because, on the interpretation given above, the function A is defined at w: indeed, at w, A takes on the value s. On the other hand if now 'w' stands for the world, then in the context of a naturalistic world view the above inference is invalid no matter how 'T' be reinterpreted, for A is no longer defined at w. In short, with this new interpretation of 'w' $A(w)$ becomes an improper description and the substitution of $A(w)$ for s is invalid because one of the conditions for the composition of functions, i.e., for the existence of the composite $T \circ A$, is no longer met. (We may, if we wish, say with Frege that '$A(w)$' now designates the null individual, but this won't save the inference.) In this regard, then, our treatment does not differ from either Russell's or Hilbert-Bernays'. The difference consists, of course, in that the existence condition is now stated separately instead of being fused with the definite description.

3.5. *Meaning Questions*

We know from Ch. 7 that meaning is contextual. Thus "The creator of the universe" is meaningful in some theodicies but not in physics, where the concept of creator does not occur. Hence we had better start by recalling the notion of a context (Ch. 2, Sec. 3.4, Definition 10). We shall reword it in terms of functions rather than predicates. *Voilà*: The ordered triple $\mathbb{C}=\langle S, \mathbb{F}, U\rangle$ is called a *context* iff S is a set of statements in which only the function constants in the function family $\mathbb{F}$ occur, and the reference class of every F in $\mathbb{F}$ is included in the universe U.

We now postulate the conditions under which a definite description, construed as a functional semistatement (Sec. 3.3), makes sense in a given context and has a referent in it.

AXIOM 9.1 Let $F(x)$ be a definite description and let $\mathbb{C}=\langle S, \mathbb{F}, U\rangle$ be a context. Then $F(x)$ *makes sense* in $\mathbb{C} =_{df} F$ is in $\mathbb{F}$.

AXIOM 9.2 Let $F(x)$ be a definite description and let $\mathbb{C}=\langle S, \mathbb{F}, U\rangle$ be a context. Then $F(x)$ *has a referent* in $\mathbb{C} =_{df} F(x)$ makes sense in $\mathbb{C}$ and F is defined at $x \in U$.

Finally we identify the referent in an unambiguous way:

DEFINITION 9.20 Let $F(x)$ be a definite description that makes sense and has some referent in a context $\mathbb{C}=\langle S, \mathbb{F}, U\rangle$. Then the value $y=F(x)$ is called the *referent* of $F(x)$.

Notice that the above axioms are postulates not definitions: indeed they do not define sense and reference but just stipulate the conditions for a definite description to have them. Now, in our semantics, for an expression to be significant it must have a (vicarious) sense. (Recall Ch. 7.) It follows that "The present king of France" and other referentially vacuous definite descriptions may be meaningful in certain contexts. In other words, an empty description may stand for a concept, i.e., something endowed with a sense, even though it fails to denote, i.e., to have a real counterpart. Consequently ⌜The present king of France is bald⌝ is meaningful as well: it has a sense – and even a referent. Whether this referent exists, i.e., whether the statement is true, is another matter – one to be considered shortly.

Our theory is at variance with the view that empty descriptions and sentences containing them "suffer from infelicity" because devoid of real reference (Austin, 1962). This opinion would be justified if meaning were identified with extension. But, as we have seen in Sec. 1 and before, such a referential view of meaning would be crippling to science, where statements about "theoretical" (i.e., hypothetical) entities have to be made, discussed and put to the test long before it becomes possible to ascertain whether they have any real referents.

Another advantage of our theory is that it dissolves a well known puzzle about "intensional expressions", i.e., non truth functional formulas. Consider the descriptions

The square root of 4. (A)
The atomic number of Helium. (B)

Since A and B have the same designatum, namely 2, they are equivalent. And, being equivalent, they should be exchangeable to a nominalist (e.g., Ajdukiewicz, 1967a). But of course the two descriptions have different referents: while A concerns 4, B is about Helium. Therefore A and B are not synonymous. Consequently, on our theory they are not everywhere exchangeable, whether in "extensional" (i.e., truth functional) or "intensional" (i.e., non truth functional) contexts. Hence the statement

Archimedes knew that the square root of 4 equals 2. (C)

is not the same as

Archimedes knew that the atomic number of Helium is 2. (D)

3.6. *Truth Questions*

Consider again the much discussed statement ⌜The present king of France is bald⌝. Can we say that it is false or that it is true? Russell held it to be false and most philosophers seem to share his view for the simple (ergo suspect) reason that it is a proposition and thus (allegedly) either true or false and, not being clearly true, it must be false. However, some philosophers have felt dissatisfied with this view. Thus Frege, and at one time Strawson (1950), held that statements containing empty descriptions are neither true nor false. More recently Strawson (1964) has come to the conclusion that his former stand is not compelling: that each side has its merits and that it is unimportant which is taken.

In our view truth and falsity do not inhere in propositions but are (sometimes) attributed to them (Ch. 8). Now, in order for a statement containing a definite description to be assigned a truth value in a given context, it must point to a definite referent in some context. For, if it fails to have a referent, then the statement cannot be "faced" to it in order to be assigned a truth value. Therefore we stipulate

AXIOM 9.3 A statement containing a definite description $F(x)$ may be assigned a truth value in the context $\mathbb{C} = \langle S, \mathbb{F}, U \rangle$ iff $F(x)$ has a referent in $\mathbb{C}$ and this referent exists.

For example, "the luminiferous ether" is a definite description whose sense may be taken to be determined by some ether theory (recall Ch. 5). In all mechanical ether theories, such as Cauchy's, the statement

$$e = \ulcorner \text{The luminiferous ether is elastic} \urcorner.$$

is not only meaningful but also true. However, since the predicate "ether" is absent from modern optics, by our Axiom 1 (Sec. 3.5) it makes no sense in it. Consequently the definite description "the luminiferous ether" makes no sense in modern optics. Ergo no statement containing this definite description makes sense in modern optics: in particular, e above is not meaningful in modern optics, or MO for short. And, not being meaningful in MO, e cannot be assigned a truth value in this context. In other words, the truth valuation function $\mathscr{V}$, that may be construed to carry the couples statement-context into truth values, is not defined for the pair $\langle e, MO \rangle$. In short, $\mathscr{V}$ has no value at $\langle e, MO \rangle$ even though $\mathscr{V}(e, E) = 1$, where 'E' is short for "ether theory". (Nor does e take the value "indeterminate", as some interpretations of many valued logic would have it. A function not determined, or not defined, for a certain value of its argument has no value at it. Depending on the theory of truth $\mathscr{V}$ may take two or more values but it cannot take the value "indeterminate".) What holds for the ether holds for contemporary French kings: in the context of contemporary history ⌜The present king of France is bald⌝ is neither true nor false. (It only presupposes a false proposition.) So let us stop arguing about it.

Finally note that the Liar paradox does not occur in our semantics, because 'What I am now saying' is a definite description not a proposition, hence it cannot even be false.

3.7. *The Real Size of the Theory of Descriptions*

The analysis of definite descriptions has been overrated to the point of having been regarded as Russell's chief contribution to philosophy – which is a way of belittling *PM*. On the other hand the theory of descriptions has been underrated and even misplaced by many ordinary language philosophers. Some took it to be a matter of literary criticism, others took it for a grammatical analysis of the definite article – as if 'my wife', 'Plato's teacher', and 'that weird chap in the corner' did not qualify as definite descriptions. As has been pointed out, all advanced languages teem with definite descriptions even if, like Latin, they lack definite articles. Also mathematics, the language of science, is shot through with definite descriptions – just recall 'the sine of 10°', 'the composition of f and g', and 'the indefinite integral of f'. So is contemporary science, which has more use for definite descriptions – e.g., in the form of spatio-temporal coordinates – than for proper names. This renders Quine's proposal of assimilating all names to definite descriptions attractive. However, (*a*) even if in practice we often do proceed this way, it is convenient to have the complex notion of description analyzed into simpler concepts, and (*b*) in metaphysics we need the concept of a non-descript or bare individual that may function as one of the building blocks of the notion of a fully qualified thing. (See Ch. 1, Vol. 3 of this *Treatise*.)

Since definite descriptions are rampant, it behooves the philosopher to analyze them. But the analysis need not introduce any new technical concept: we have seen that descriptors are reducible to standard components of elementary logic and mathematics. Our evaluation of description theory lies thus midway between the two views that are currently dominant: instead of either renouncing definite descriptions or inflating them, we hold them to be normal constituents of any language with a reasonable expressive power. Moreover in our view the syntax of definite descriptions is trivial: only their semantics is somewhat complex, in the sense that it involves the notions of sense, reference, and truth. But then this is what semantics is all about: sense, reference, and truth.

This is as far as we go in applying our basic doctrines to issues in pure semantics. The next chapter, which is also the last, explores some of the relations among semantics and its neighbors.

CHAPTER 10

NEIGHBORS

In this, the final chapter, we shall peek at some of the adjacent fields of inquiry in order to better locate ours. We shall try and see what they look like in the light of our semantic candle. In each case we shall have to limit ourselves to examining a few typical problems. Moreover our discussion will be rather quick, as our aim is to explore the nature of the bonds between the semantics of science and its nearest neighbors rather than to examine the latter in detail. We shall first glance at mathematics, or rather its philosophy, then at three traditional branches of philosophy: logic, epistemology, and metaphysics.

1. Mathematics

1.1. *The Relevance of Semantics to Mathematics*

That mathematics is relevant to exact semantics, is analytically true, for exact semantics is nothing but semantics built *more geometrico*. The question is whether basic semantics is relevant to mathematics – whether the former can add anything to the semantics of mathematics, or model theory. And this is not obvious.

Consider the notions of designation, reference, sense, and truth – all of which would seem to be of interest to mathematics. (The concept of model, in the model theoretic sense, is the exclusive property of the semantics of mathematics and it does not seem portable to other fields: recall Ch. 6, Sec. 2.4.) The concept of designation is almost trivial and so is that of reference in the case of mathematics – though not so in relation to science. Indeed, to all but the literalists it is clear that mathematical symbols designate mathematical constructs. It is equally plain that mathematical concepts and statements either are or concern mathematical objects, which are so many constructs. Thus while "two" does not refer to anything, "even" refers to integers and ⌜Two is an even number⌝ refers to two, which can in turn be designated by the numeral '2'. So far nothing shattering.

As to the third generic semantic concept, namely that of sense, we do have something to say about it, particularly since model theory deals only in extensions. (Recall Ch. 6, Sec. 2.3.) But in our view the full sense of a theory, whether mathematical or factual, is ultimately determined by the postulates of the theory. (See Ch. 5, Sec. 5.) Hence we have nothing to add to what was said in Chapters 4 to 7. (For specifics concerning meaning in mathematics see Castonguay (1972).) However, we may caution against the persuasive or ideological uses of 'sense' and 'meaning' as exemplified by the war-cry that whatever fails to be constructive, such as the axiom of choice, is meaningless (cf. Lorenzen, 1967). This Principle of Intolerance is meaningless unless backed by a definite theory of sense (or of meaning) in the spirit of constructivism – which, alas, is not available. So much for sense in mathematics.

On the other hand we have little to say about the fourth capital term of semantics, namely 'truth', with reference to mathematics. The reason is that, although 'truth' parades as a general term, it is not: 'truth' is an ambiguous term designating two radically distinct concepts, namely those of formal truth and factual truth. Whereas factual truth is elucidated in terms of external reference and empirical evidence, formal truth is elucidated in terms of satisfiability and proof. (From a pragmatic point of view proof has the upper hand, since showing that a formula is true in a certain model, or under a certain interpretation, boils down to proving the formula in the theory of the model.) Whereas factual truth is the object of a special theory, formal truth is elucidated within model theory, which is a multiple purpose theory. (Also the notion of potential formal truth is definable in model theoretic terms: see Robinson, 1965. And even the notion of partial formal truth can be handled by model theory: see Chang and Keisler (1966).) Because our system of semantics is geared to factual science, it is as irrelevant to mathematical truth as the model theoretic concept of truth is alien to factual truth.

In sum, only our theory of sense (Chapters 4 and 5) makes sense with reference to mathematics. Having nothing further to offer we shall close with a couple of critical remarks.

1.2. *On Extensionalism*

'Extensionalism' is sometimes taken to designate the thesis that ordinary logic is the only logic one needs – that, in particular, modal logics are dispensable. We accept this thesis but reject that designation as mistaken and misleading. (See Ch. 4, Sec. 1.3.) The authentic extensionalist thesis is, in a nutshell, that *every concept worth its name is a set*. When adopted in semantics, extensionalism effects a hecatomb: it kills sense (in every sense) and it is apt to conflate reference with extension.

The prestige of the extensionalist thesis stems from the belief that it has subdued mathematics. In fact it is widely held that (*a*) set theory is fully extensional and (*b*) the whole of mathematics is reducible to set theory. However, these two tenets are, to say the least, controversial. First, set theory contains a basic concept, that of membership, which is not construed as a set but as a relation between something (whether individual or not) and a set. Not only is the membership relation not defined as a set (of ordered pairs), but the latter is partially defined in terms of the former, namely by saying that, if $x \in y$, then y is a set. (But of course only the totality of postulates of some set theory does the job of determining the full sense of $\in$.)

Second, set theory contains a postulate, the principle of separation (or its forerunner the principle of abstraction), that relates predicates to sets – the extensions of the former – without defining the former in terms of the latter. (Recall Ch. 4, Sec. 1.2 and Ch. 9, Sec. 1.1.) If this were the sole assumption of every set theory then it would support Russell's view that reasoning on properties is primary – and by the same token it would disqualify Bourbaki's Olympian contempt for the allegedly anachronistic *raisonnement en compréhension*. However, a balance between the two extremes looks more realistic than either.

Third, although nowadays nearly every mathematical theory – much to Wittgenstein's chagrin – uses set theoretic concepts and even some formulas of set theory, it must add something of its own if it is to get off the ground. Without such specific concepts and assumptions, that are not reducible to (definable in or deducible from) set theory, there would be no mathematical theories other than set theory. (The fact that most new mathematical concepts can be characterized with the *help* of set theoretic concepts does not entail that they were already "contained" in set theory. Likewise an organism is not prefigured in its physical components.) Thus the definition of one of the simplest mathematical structures, namely a semigroup, requires the notion of associativity, which set theory does not define.

The upshot of the preceding discussion, conjoined with that in Ch. 4, Sec. 1.2, is that the extensionalist thesis is false in mathematics. On the other hand the program of using set theoretic concepts in the whole of mathematics and its applications has been tremendously fruitful although it may not be the last word. In any case "set theorification" should be kept distinct from extensionalization.

1.3. *On Objectivity*

Our second and last remark will concern objectivity. It is almost universally agreed that mathematics is objective – but this because everyone seems to have his own concept of objectivity. In any event this much seems to be true: Although mathematics is a creature of the brain, it is not in the same category with (wild) dreams and (silly) fairy tales. It does not hold by arbitrary decree or because I want to or because we happen to believe in it. Once born, a piece of mathematics ceases to be subjective and acquires a certain objectivity. In fact it is as objective as factual science – only in a different sense. Let us compare these two modes of objectivity.

The objectivity of factual science consists in its reference to external objects alone (semantic objectivity) and in its impersonal or publicly scrutable checking procedures (methodological objectivity). Even mental processes, when studied by scientific psychology, are handled as external objects and in a publicly criticizable way. Mathematics, on the other hand, fails to refer to external objects, whether ideal or material, hence it is not semantically objective. Nor is it semantically subjective: it is not concerned with our private mental states. The objective/subjective dichotomy, in the semantic sense, applies to mathematics no more than the hot/cold dichotomy does.

But mathematics is *methodologically objective*, though not in the sense that it employs empirical testing procedures (observation, measurement, experiment). The methodological objectivity of mathematics consists in (*a*) impersonality, (*b*) abidance by the assumptions and rules agreed on beforehand – including the general principles of rational argument, and (*c*) the justification of both assumptions and rules in terms of impersonal values such as coverage, systemicity, validity, and clarity. All three traits are shared with factual science, which adds empirical testing to them. Hence mathematical objectivity is but a special case of scientific objectivity: objectivity without objects other than mathematical constructs.

In sum, mathematics does not possess the semantic objectivity that both Platonists and vulgar materialists assign it: mathematics is neither about self existent ideas hovering over the world nor about the latter. Mathematics is methodologically objective in the sense that its procedures are exoteric. But mathematics is not semantically objective. To claim that it is (as Popper (1972), (1974)) does, i.e. to hold that mathematics is just as objective as physics, is to conflate epistemological realism with objective idealism.

2. Logic

2.1. *Analyticity*

The problem of analyticity has held the attention of philosophical semanticists to the detriment of all other semantical problems, thus creating a serious imbalance and causing boredom. So far in this work we have made a modest use of the concept of analyticity, or rather one of them, without elucidating it. The time is ripe for such an elucidation. As usual we must choose the target before shooting: we must decide whether we want to have a narrow concept or a wide one. Having tried a concept of analyticity of maximal extension in the past (Bunge, 1961b) we shall now choose the most restricted one. The strict (and relative) acceptation of 'analytic' is this: An analytic formula is one that either holds under all interpretations (in all models) or consists in a definition. More explicitly, we adopt the following

DEFINITION 10.1 A formula ϕ in a theory T is *analytic in* $T =_{df} \phi$ is either a definition in T or model free.

This convention does not employ the notions of logical form and meaning but it fits the statements that are (formally) true "by virtue of their form", i.e. the tautologies, as well as those that hold "by virtue of the meanings of their parts" – e.g. the dictionary definitions (which Carnap called 'semantic postulates').

To call 'synthetic' all those formulas that are not analytic would be misleading in view that 'synthetic' has so often been equated with 'having a factual content' or with 'empirically testable' or with 'informative'. The extralogical formulas of a mathematical theory are not analytic in the sense of Definition 1 but it would be queer to call them 'synthetic': they are *neither* analytic nor synthetic (in the sense of empirical or factual). In other words, *the analytic/synthetic distinction is not a dichotomy*. We must distinguish more than two species of statement, at least the following ones:

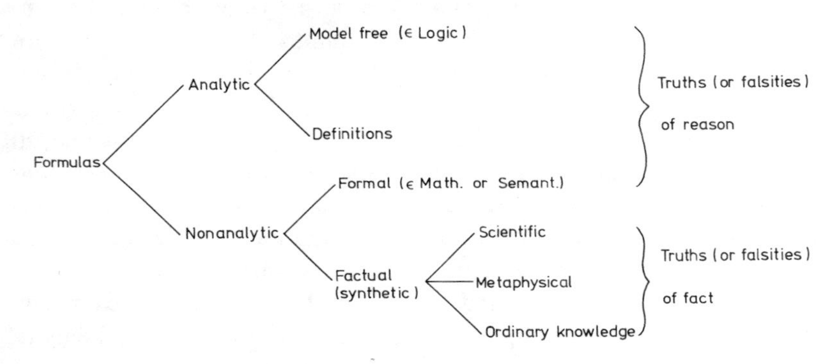

The preceding definition of "analyticity" seems to be precise and clear but it does not solve the practical problem of spotting analyticity in every single case. Such a problem will come up in ill organized conceptual bodies such as those of ordinary knowledge and of intuitively formulated scientific theories. For example, the statement ⌜Seeds germinate when they fall on fertile soil⌝ may be taken either as a lawlike statement or as a covert (and partial) definition of soil fertility. If ambiguities like this arise in the flabby context of ordinary knowledge, so much the worse for the latter – or for the attempt to draw technical distinctions in nontechnical contexts. No such status ambiguities occur in a reasonably well organized conceptual body. What does happen in the latter case is that one and the same formula may be analytic in one systematization and nonanalytic in the next – but we made room for such switches by relativizing the concept of analyticity to that of theory in Definition 1.

In conclusion, our definition of analyticity solves the theoretical problem of elucidating this notion but is not a fool proof *criterion* for telling whether a given formula is analytic – much less for identifying analytic components in open contexts. (Likewise a definition of consistency is not enough to prove the consistency of any particular theory.) Nevertheless a difficulty in drawing a distinction in particular cases does not prove that there *is* no difference. Nor does it prove that it is impossible to reorganize the body in question so as to bring out the difference, not just between analytic and nonanalytic statements, but also between hypotheses and conclusions, and so on.

If the preceding considerations are substantially correct then the concept of analyticity is far less important to philosophy than that of truth of reason or necessary truth (Leibniz, 1714). While this wider concept serves to distinguish formal science from factual science, the narrower one of analyticity serves to characterize logic *vis-à-vis* the rest and, in particular, mathematics. Indeed, while the two formal disciplines contain only necessary truths, logic has the monopoly of analytic formulas other than (extralogical) definitions. Mathematics does not get started unless some nonanalytic (and nonfactual) formulas, containing extralogical concepts such as "$\leqslant$" and "$+$", are added to a basic logical broth. For example, the theory of groupoids adds the following ingredients to the logical predicates: an abstract set and a binary operation on it.

In sum, given a set R of rules of inference, logic is self-generating, i.e. the totality of analytic formulas follows from no assumption whatever. (The catch is in R.) On the other hand a piece of mathematics requires, in addition, a nonempty set of extralogical (but also nonfactual) assumptions. In symbols: while $\emptyset \vdash_R \mathit{Logic}$, $A \vdash_R \mathit{Mathematics}$, where A is the set of mathematical assumptions. (A change in R may shift the border line without obliterating it.) An equivalent characterization of the difference is this: While the truths of logic are satisfied in all models, those of mathematics are satisfiable only in some models – sometimes just one, at other times infinitely many, but never all.

We uphold then the analytic/synthetic distinction, indicted in recent times (Quine, 1952). However, we do not define analyticity in terms of the information needed for understanding a sentence, this being a pragmatic not a semantic concept of analyticity; hence we are not bothered by Quine's examples. Furthermore we do not regard the distinction as a dichotomy or as being central to the whole of semantics and the philosophy of formal science. The essential distinction, as far as epistemology and the philosophy of science are concerned, is that between truths (or falsities) of reason and truths (or falsities) of fact. Were we to give it up we might be tempted to indulge in either of the following oddities: empiricism with regard to formal science, and apriorism with reference to factual knowledge. To show that these are not imaginary perils we quote an instance of each. The empiricist philosophy of mathematics is defended by no less than Kalmár (1967 criticized by Goodstein 1969). And the converse move, of eliminating extralogical postulates in favor of definitions, is not only a favorite ploy with conventionalists (criticized by Enriques, 1943, pp. 250–251) but has also been tried by Quine and Goodman (1940, Quine, 1964, criticized by Bunge, 1967a, I, pp. 132–133). Finally, many mechanics textbooks contain traces of Mach's attempt to blend the two moves – taking law statements for definitions and conversely (see Bunge, 1966). If only to avoid such errors it is imperative to hold the fort – not the analytic/synthetic nondichotomy but the rational/factual dichotomy. Still, it is equally imperative (*a*) not to insist on drawing such metatheoretical distinctions with regard to conceptual bodies that have not been given a deductive structure, and (*b*) not to forget that pure reason is an invention of certain organisms.

We conclude by sketching an enlarged notion of semantic analyticity, or analyticity by virtue of meaning. This notion is suggested by our incursion into the topology of the intension space and, in particular, by our elucidation of Wittgenstein's coarse notion of family resemblance (Ch. 4, Sec. 2.5). The idea is that an exact tautology can be approximated, arbitrarily closely, by a *quasitautology* or statement that is *almost* completely true (formally). Consider ⌜All *A*'s are *A*'s⌝ and replace one of the occurrences of *A* by *B*, where *B* is a close relative of *A* in the sense that *B* belongs to a small neighborhood of *A* in accordance with Definition 9 in Ch. 4, Sec. 2.5. Then both ⌜All *A*'s are *B*'s⌝ and ⌜All *B*'s are *A*'s⌝ will be *quasitautologies*. The process is reversible: given a nontautologous statement, Theorem 4.11 in Ch. 4 allows us to construct a whole sequence of propositions the limit of which will be a tautology. Take for instance the much debated statement ⌜The fittest organisms survive best⌝. Replace "fittest" by "most fertile" (in accordance with the Pickwickian acceptation adopted by population genetics) and "survive" by "propagate" (or "have the best reproductive survival") and you get an exact tautology. In sum, if meaning is a matter of degree so is analyticity. And, just as in the case of the multiplicity of truth values, the continuum of degrees of analyticity is consistent with rigid two valued logic.

2.2. *Definition*

The concept of a definition has proved more problematic than it deserves (see e.g. Bunge, 1967a, I, pp. 117–139). This is partly due to the adoption of too wide a construal of 'definition' – one that makes room for practically every determination of one group of constructs by another. A number of problems drop out if we adopt a narrower concept – just as with the case of analyticity. The narrowest of all, and that which we favor, is Peano's construal of every definition as an equality of the form: "*the defined = the object defining*", where the object in question is either a sign or a construct (Peano, 1921).

Example 1 The definition of set equality: ⌜If A and B are sets, then $A = B =_{df}$ The membership of A = The membership of B⌝. *Example 2* The definition of "$\leqslant$" in the relational structure $\mathcal{N} = \langle N, + \rangle$, where N is the set of natural numbers and $+$ the addition operation: ⌜If x, y and z are natural numbers, then: $x \leqslant y =_{df} (\exists z)\,(x + z = y)$⌝. *Example 3* The axiomatic definition of a groupoid: ⌜The structure $\mathcal{G} = \langle G, \cdot \rangle$, where G is a nonempty set and $\cdot$ a binary operation on G, is a *groupoid* $=_{df}$ For all x and y in G, $x \cdot y$ is in G⌝. *Example 4* The definition of the energy density of an electric field: ⌜If E represents the intensity of the electric electric field f, then $U(f) =_{df} (1/8\pi)\, E^2$⌝.

Note the following characteristics of a definition, whether explicit or implicit, conditional or unconditional, by abstraction (as Example 1) or axiomatic (as Example 3) or of any other type provided it is centered on the equality concept. Firstly, it may be taken to define a *construct or the symbol* chosen to designate it. One does not define things other than constructs or their symbols. For example one does not define light (but describes it); rather, one defines some concept of light (or the terms 'light', 'lux', 'lumière', etc.). More on this toward the end of this section. Secondly, every definition is *relative* to some context, whether a relational structure (as in Example 2) or a full theory (see Padoa, 1901 and Tarski, 1934). The contextual or relative character of definitions should be kept in mind in order to avoid some of the mistakes mentioned at the end of the preceding subsection and if one wishes to understand why one and the same word, when defined in different contexts, may acquire different significations – i.e. may designate different constructs. Thirdly, the concept "$=_{df}$" of equality by definition should be assigned all the formal properties of *equality* (or congruence), in particular the symmetry property, in order to ensure the exchangeability of definiens and definiendum. This requirement is less trivial than it looks for, from a methodological point of view, there is no such symmetry: the definiendum and the definiens cannot exchange places. In other words, from a pragmatic point of view '$=_{df}$' is a metalinguistic symbol that intends to convey the idea that the left hand side is determined by (is a function of) the right hand side not conversely. (This is how definitions were treated in *Principia Mathematica*.) However, this difference between definiens and definiendum is metatheoretical and methodological not semantical, for it points to a difference in status and role not in meaning. Semantically speaking, to define A as B is to equate A with B. So much so that a justification of a definition may consist in proving the identity of definiens and definiendum: no such proof would be needed if every definition were just a linguistic convention.

The narrow construal of definitions as equalities has the following advantages. First, the identity in meaning (sense and reference) of definiendum and definiens is brought out. (On the other hand an equivalence does not warrant meaning identity since equivalents, though coextensive, need not be cointensive. More on this in a moment.) Second, every definition of the form $\ulcorner A =_{df} B \urcorner$, with A and B propositions, entails the equivalence $\ulcorner A$ iff $B \urcorner$ (or $\ulcorner$For all x, Ax iff $Bx \urcorner$) but not conversely. (If A and B are identical then either can take the place of "the other" in the tautology $\ulcorner A$ iff $A \urcorner$.) Third, every formally correct definition belongs by right in the class of analytic formulas – by Definition 1 in Sec. 2.1. Consequently the old dispute as to whether a convention might be true is settled: All definitional equalities are necessary truths, albeit rather cheap ones. (In fact they are the only necessary truths found in factual science apart from the formal truths employed in inference.) Fourth, any sentence expressing a definitional equality belongs to the object language of the conceptual system concerned: no further language level need be invoked since '$=_{df}$' is no longer a metalinguistic sign but just a metatheoretical status symbol. Fifth, proofs of concept independence may often be simplified, as identities are usually easier to check than equivalences. Moreover instead of employing semantic techniques (e.g. Padoa's) for checking concept independence (or undefinability) in a context one may try the following alternative procedure. Check whether definiens and definiendum meet the necessary condition of coextensiveness: if the extensions do not coincide then the suspected definiens will differ from the definiendum. However, if the *LHS* and the *RHS* are coextensive nothing can be concluded: the technique is one for disproving not for proving concept independence.

In the literature several alternative forms are used or even prescribed for definitions, chiefly the following:

'*A*' names (designates) *B* (*D*)
"*A*" means the same as "*B*" (*S*)
A iff *B*. (*E*)

However, each of these formats has some disadvantage from which that of equality is free. In fact *D* multiplies unnecessarily the levels of language and does not guarantee substitutivity. *S* is not applicable prior to interpretation (Padoa, 1901) and moreover it presupposes either an intuitive (muddled) concept of meaning or a universally accepted theory of meaning – which, alas, is not forthcoming. Moreover if *S* is adopted then it prevents us from defining "meaning" without circularity. In any case, notwithstanding a popular view, a definition should not be required to "explain the meaning" of the definiendum – except of course in a pragmatic sense. Otherwise implicit definitions should be unacceptable. Consider, if in doubt, the recursive definition of the addition operation in the relational structure $\langle N, S \rangle$, viz.,

$$x+0=x \quad \text{and} \quad x+Sy=S(x+y) \qquad Df$$

where 'Sx' designates the successor of x. Since "+" appears on both sides, it is neither explained in terms of the previous concepts nor eliminated in favor of them (Goodstein, 1968).

Finally, the equivalence format *E* is open to the following objections. First, equivalents, though coextensive, are not necessarily cointensive – hence they may not be substituted for each other without a concomitant alteration in meaning. Second, it is important to keep the difference between equivalence and equality. Thus the biconditional ⌜*A* is provable iff *A* is a tautology⌝ is a metalogical theorem not a definition of provability. Likewise Tarski's principle ⌜"*s*" is true iff *s*⌝ is a biconditional not a partial definition of truth.

Finally let us emphasize that only signs and their designata can be defined (provided they are not primitive). Factual items can be described, explained or predicted: not being constructs, facts cannot be built out of constructs. In sum there are no "real definitions". Hence, *pace* Suppes (1967, 1969) and Freudenthal (1970, 1971), the technique of axiomatic definition, ideal as it is for characterizing mathematical objects, cannot be expected to define the concrete objects described (not defined) by scientific theories (see Salt (1971) and Bunge (1973b)). For example, the equations of mechanics define (implicitly) a concept of body not bodies. So much for the misprint 'axiomagic'.

Like other methodological categories, that of definition has several dimensions: linguistically it is usually an abbreviation; epistemologically it is a way of building new concepts out of old; pragmatically it is a heuristic device and often a time saving device. We have dealt only with the semantic aspect of definition and cannot go any further.

2.3. *Presupposition*

A presupposition is a tacit or undisclosed supposition – one that can and should be brought to light by analysis. We shall distinguish six concepts of presupposition, all of them relevant to the analysis of scientific knowledge. They shall be introduced by so many definitions. In each of these 'A' and 'B' will designate sets of formulas.

First the *proof theoretic* notion of presupposition:

DEFINITION 10.2 *B presupposes A relative to the proof* $P =_{df} A$ belongs to the set of premises occurring in the proof P of B.

Change the proof and B may cease to presuppose A.

Next a first *semantic* concept of presupposition:

DEFINITION 10.3 *B weakly presupposes A meaning-wise* $=_{df} A$ is sufficient to determine the meaning of B.

For example, if a scientific theory B involves the concept of time then B must weakly presuppose, *inter alia*, some theory A of time. (The whole lot of theories presupposed meaning-wise by any given theory may be called the *background* of the theory: Bunge, 1967b.) Any number of alternative presuppositions may do the job, though not all of them equally well. Hence no single presupposition of this kind may be necessary.

For example, Newtonian mechanics can be axiomatized by presupposing either absolute space and time or relational space and time.

DEFINITION 10.4 *B strongly presupposes A meaning-wise* $=_{df}$ *A* is necessary to determine the meaning of *B*.

For example, arithmetic is a strong presupposition of any quantitative theory, both meaning-wise and proof theoretically.

Let us now distinguish two concepts of *alethic* presupposition:

DEFINITION 10.5 *B weakly presupposes A truth-wise* $=_{df}$ The truth of *A* is sufficient for the truth of *B*.

In other words, *B* is a weak alethic presupposition of *A* just in case $A \models B$. Consequently this concept of presupposition belongs to model theory. Therefore it has only an indirect relation to factual science, namely via mathematics. In factual science the most we can do is to check whether *B*, or rather some consequences of *B* conjoined with some extra premises (e.g. data), are approximately true. To put it otherwise: In factual science the most useful concept of entailment is the syntactic ($\vdash$) not the semantic ($\models$) one, since we must feel free to investigate the logical consequences of any assumption before assigning the latter any truth values. (Note that, although $\vdash$ and $\models$ are coextensive, they fail to be cointensive.) Next comes

DEFINITION 10.6 *B strongly presupposes A truth-wise* $=_{df}$ The truth of *A* is necessary and sufficient for the truth of *B*.

If weak alethic presupposition has a remote relation to factual science, strong alethic presupposition is even further removed. Finally we have the *methodological* concept of presupposition:

DEFINITION 10.7 *B presupposes A methodologically* $=_{df}$ The test of *B* uses *A* without questioning *A*.

For example, although electromagnetism is a fundamental theory in the sense that it can be formulated without resorting to any other scientific theory, its empirical tests presuppose a number of other theories – in fact as many as needed to design and control the instruments employed in the tests. Operationism could easily muddle up this issue by claiming that, this being so, then every theory presupposes every other theory. As a matter of fact the so called Copenhagen interpretation of quantum mechanics claims that, although quantum mechanics retrieves classical mechanics without being fully consistent with it, it also presupposes the latter because the former theory must be interpreted in terms of experiments, and every experiment must be accounted for by classical physics (Landau and Lifshitz, 1958, p. 3, criticized by Bunge, 1970b, pp. 310 ff.).

This winds up our examination of the (tiny) impact of our semantics on logical theory.

3. Epistemology

3.1. *The Status of Epistemology*

The collection of opinions concerning human knowledge, i.e., epistemology, used to be the nucleus of modern philosophy. Over the past century three other lines of inquiry have come to dispute the territory of epistemology: biology (following Helmholtz and Mach), psychology (following Piaget), and semantics (after Tarski and Carnap) – not to speak of the philosophy of science, which overlaps with epistemology, and pragmatics, which is still in the foetal state. Each of those disciplines offers persuasive arguments for absorbing the whole of epistemology. Let us weigh them.

The argument for surrendering epistemology to biology and psychology seems compelling: Perception and ideation are but so many aspects of man's effort to adapt himself to, as well as alter, his environment. Hence they belong to the study of man the animal. More particularly since sensing, perceiving, representing, and inferring are functions of the central nervous system, the study of knowledge falls within the province of neurophysiology and psychology.

All this is quite true. It has become ludicrous to speculate about perceiving without regard to the psychology of perception, about knowing without looking into learning theory, and so on. Some of the traditional problems of epistemology have indeed been snatched away by psychology. Epistemology is thus becoming biologized as demanded by Campbell (1959). This process is irreversible despite the efforts of philosophical psychology, which can now thrive only on the shortcomings of scientific psychology. Still, philosophy has a right to investigate knowledge (and for that matter anything) provided it does so from a different angle, with its distinctive means and aims, and as long as it learns from scientific psychology.

The philosophical study of knowledge includes a theoretical examination of the following subjects: (*a*) general nature of man's knowledge of his environment and of himself – i.e. knowledge as a metaphysical subject (in N. Hartmann's words, a subject of the *Metaphysik der Erkenntnis*); (*b*) kinds of knowledge (intuitive, rational, etc.) and their interrelationships; (*c*) relation between physical object statements and sense data statements; (*d*) concepts of factual truth, error, and error correction; (*e*) traditional dichotomies such as subjective-objective, a priori-a posteriori, experiential-conceptual, and intuitive-rational. None of these themes is claimed by biology or psychology – nor can they be studied seriously without the help of these disciplines.

As to the credentials of semantics, they too look impressive: while traditional epistemology dealt with truth in a metaphorical way and neglected meaning altogether, semantics offers exact theories of each of them. This is true, but still it may rejoined (*a*) that those exact theories are not concerned with factual knowledge, which is the main theme of epistemology, and (*b*) semantics pays no attention to the problems listed above. So, epistemology does have a territory of its own *vis-à-vis* semantics. Moreover, it could be argued that semantics, or at least the semantics of factual knowledge, is but *a part of epistemology* – namely that portion concerned with the reference, sense, and adequacy of human knowledge in general as different from the cognitive process.

Whatever view on the semantics-epistemology relations one may adopt, two things seem certain. The first is, that the two fields overlap. The second, that it matters little whether there is a border line (rather than a stripe) between the two as long as something be done about the problems themselves. Let us pick two representative problems of epistemology and see what our semantics has to say about them.

3.2. *Representation vs. Instrument and Picture*

One of the dichotomies tackled by epistemology is that of *a priori* vs. *a posteriori* knowledge. Because 'a priori' is usually taken to signify 'prior to experience' and because 'prior' is an ambiguous word, 'a priori knowledge' can be a misleading expression. Thus while it is true that logic *holds* irrespective of experience, it is false that it *originated* independently of experience: logical investigation is just a fragment of human experience. For this reason – i.e., to avoid the genetic fallacy – it is advisable to draw a clear distinction between origin and validity, and to relativize the notion of *a priori* to a body of knowledge. The two requirements are met by the following conventions.

DEFINITION 10.8 The set S of statements is *a priori* relative to the body of knowledge $K =_{df}$ No member of S presupposes any member of K either meaning-wise or truth-wise.

DEFINITION 10.9 The set S of statements is *absolutely a priori* $=_{df}$ S is *a priori* relative to every body of knowledge.

In this sense logic is absolutely a priori, mathematics is a posteriori (= not a priori) with respect to logic, certain branches of mathematics are a posteriori relative to others, science is a posteriori relative to the whole of formal science, and certain branches of science are a priori relative to others. These epistemological notions of a priori and a posteriori are not the same as the psychological and the methodological concepts similar to them. For example, a new conjecture in logic is a priori and moreover absolutely so, but it is methodologically a posteriori in the sense that it must be checked before being included in logic. And a new scientific hypothesis, though a posteriori in the epistemological (or semantical) sense, is psychologically a priori in the sense that it comes to us before any new data.

Whatever bit of human knowledge is prior to some other bit may be said to function as an *instrument* for the latter. The epistemological questions are (*a*) whether the whole of factual science is but an instrument for action – as pragmatism and the anti-science movement hold, and (*b*) whether every scientific theory is nothing but a data processing instrument (conventionalism, nominalism, pragmatism, and computerism). That factual knowledge can be used as an instrument, for good or for evil, is beyond dispute: the question is whether, in addition, it represents reality (or at least experience) and if so how.

There are, of course, as many answers to the above question as there are epistemological schools. The most popular are direct (naive) realism and classical empiricism, according to which sensations and ideas, even if logical or mathematical, represent and moreover reflect or copy factual items. According to critical realism, on the other hand, our perceptual and conceptual representations of external objects are "only *signs* [symptoms] of external objects, and in no sense *images* of any degree of resemblance. An image must, in certain respects, be *analogous* to the original object (...) For a *sign* it is sufficient that it become apparent as often as the occurrence to be depicted makes its appearance, the conformity between them being restricted to their presenting themselves simultaneously" (Helmholtz, 1873, p. 391). This is, roughly, the view we adopted in Chs. 2 and 3: factual knowledge *refers* to external objects and *represents* them but it is *symbolic* rather than pictorial. Moreover the representation it effects is one-sided rather than full, and *global* (whole theory-factual domain) rather than pointwise. (Cf. Ch. 6, Sec. 3.4.) How else are we to explain that scientific theories so often go wrong and that they do not depict anything although they explain and predict?

Theoretical concepts and scientific theories do not *picture* physical objects, let alone pointwise. (Cf. Ch. 3, Sec. 1.) They cannot picture because they are constructs. Of course they may represent and, as long as they are factual, they all refer to supposedly real items. However, only some of the elements of a theory represent: others, like lagrangians and partition functions, discharge no representational function even if they do refer. (Recall Ch. 3, Sec. 1.1, and Ch. 7, Sec. 3.4.) Conversely there are real traits, such as individual idiosyncrasies, that no scientific theory will capture unless it happens to be a theory of an individual. Even so, no scientific theory attempts to provide a complete description of its referents. (Consequently Bohr and Heisenberg were wrong in claiming that quantum mechanics provides a complete description, and Einstein was right in holding that it does not. But Einstein was misguided in his search for an alternative theory that would supply "a complete description of reality". It is not only that a description requires empirical data in addition to theories if it is to describe individual things: both the history of science and an analysis of the way theories are constructed suggest that the process of taking in more and more factual items is never ending, as stressed by dialectical materialists: see Lenin (1909).)

Furthermore theoretical concepts and theories are not abstracted from sense experience, not even from scientific experiments, so they should not be placed in the same bag with empirical concepts. For one thing they need not have the same referents. For another they never have the same sense: for, if they were cointensive, then the theory concerned would be pointless. Of course there is some correspondence between concepts and percepts, but (*a*) it is not one to one and (*b*) it is symbolic or indirect (Einstein, 1936, p. 353). Materialists need not panic: the view that our representations of the external world are symbolic rather than cinematographic is perfectly consistent with the theses that there are no autonomous ideas and that our ideas, if they happen to concern autonomous external objects, sometimes do succeed in representing the latter in a more or less adequate way. In any event the point we wished to make is that critical realism is more adequate than the picture view of knowledge, which has not gone beyond the metaphor stage anyway. Nonetheless it must be owned that the symbol doctrine, proposed one century ago, is in dire need of elaboration. We are still far away from the systematic realism demanded by Hooker (1974).

3.3. *Objectivity vs. Subjectivity*

A representation, by definition, represents something: there can be no representation in and by itself, or *Vorstellung an sich*. In particular, a conceptual representation of a factual item x is a construct y that maps (some of) the features of the object x. A representation is not a part, let alone the whole, of its object. So much for the semantics of the concept of representation – some of the complexities of which were examined in Ch. 3. In epistemological terms: "Unless the content of knowledge is recognized to have a condition independent of the mind, the peculiar significance of knowledge is likely to be lost. For the purpose of knowledge is to be true to something which is beyond it" (Lewis, 1929, p. 192). This is, of course, the realist or objectivist thesis.

The subjectivist, on the other hand, holds that the world is our representation or part of the latter. Thus Goodman claims that there is no such thing as *the* way the world is: the world is as many ways as it can be truly described, seen, pictured, etc. (Goodman, 1960). Further: "That nature imitates art is too timid a dictum. Nature is a product of art and discourse" (Goodman, 1968, p. 33). The rationale is that representations are creations not copies. True enough, but the point is that a representation, however creative, does not create its object. To claim that it does is to muddle things up (as Frege (1894) noted) and to make a mockery of the age long striving for objectivity, that supreme accomplishment of the creative scientist. For it is a declared aim of scientific research to obtain objective (impersonal and publicly testable) representations of the world. This is why scientists keep checking them and trying to improve on them. This is why any subjective elements, when spotted, are weeded out. And because of this need for keeping the distinction between objectivity and subjectivity, it behooves the semanticist to clarify it.

Whether or not a representation is objective is a matter for semantics and methodology, since a construct is said to be semantically objective if it concerns external objects alone and can be subjected to impersonal tests. (But of course those objects may be persons and they may even prove to be nonexistent.) On the other hand a construct *a* is a *subjective representation* of an object *b* just in case *a* refers not only to *b* but also to the subject *c* responsible for *a*. If this subject dependence is genuine then it must show up upon substituting a different subject for the original one. This is how it can be shown that quantum mechanics, reputedly a subjectivistic theory, is perfectly objective – namely by analyzing the referents of the basic predicates of the theory and proving the cognitive subject to be absent from them (Bunge, 1967e, 1973b).

A methodological examination should confirm or infirm the results of a semantical test for the objectivity of a representation. That examination may consist either in the control of the suspected variables or in some other empirical procedure. Consider a sequence of events of a kind assumed to be mutually independent. On a subjectivist interpretation the events should occur only when observations are actually performed – e.g. every minute. Consequently the distribution in time will be binomial or similar. On an objectivist interpretation the events will occur whether or not they are observed. In particular it may happen that the process goes on continuously over a time interval. Hence the distribution could be Poisson's or similar. We have then two rival distributions of elements among the various accessible states, i.e. two different conceptual representations or models. The conflict can be resolved by sampling. Needless to say, the problem hardly arises in the physical sciences. But it does arise in psychology, and in any case it is philosophically important to know that the problem of objectivity can in principle be solved by theoretical modeling and subsequent empirical tests.

We take it then that science espouses, albeit tacitly, a realist epistemology. Moreover science confirms critical (or indirect) realism rather than naive (or direct) realism, as shown by its distrust of the "immediately given" and its relentless effort to improve on every theoretical representation of facts. In order to better appreciable this point we list the main theses of these two versions of realism. (Cf. Bunge, 1973a.)

Direct (naive) realism	*Indirect (critical) realism*
$\mathscr{DR}$ 1 Everything is knowable.	$\mathscr{CR}$ 1 Many things and facts are knowable. Many others are not – e.g., extinct things that have left no perceptible traces.
$\mathscr{DR}$ 2 Physical objects are perceived directly: the eye is innocent.	$\mathscr{CR}$ 2 Perception can be improved or distorted by preconception (superstition, hypothesis, etc.): the eye is biased – for better or for worse.
$\mathscr{DR}$ 3 All objects are conceived directly.	$\mathscr{CR}$ 3 Constructs are formed within a body of background knowledge, unevenly true and largely social.
$\mathscr{DR}$ 4 Every conceptual representation is at least homomorphic.	$\mathscr{CR}$ 4 Conceptual representations are rarely pointlike (e.g. homomorphic), mostly global (whole theory-whole factual domain).
$\mathscr{DR}$ 5 All sensations and conceptions mirror real objects.	$\mathscr{CR}$ 5 Some constructs are representational while others are purely syntactic. And many hypotheses and theories intended to represent real factual items turn out to be wholly fictitious.
$\mathscr{DR}$ 6 Perfect knowledge (complete and fully true) is possible.	$\mathscr{CR}$ 6 Factual knowledge is always imperfect (incomplete and at best partially true). But it is perfectible.

Note the following points. First, although realism (whether direct or indirect) presupposes the ontological hypothesis that there are things in themselves, i.e., objects existing independently of any mind, it does not espouse materialism. Indeed a realist need not entertain the ontological hypothesis that every existent is material. (Nor does materialism entail realism: a materialist need not believe that matter can be known: witness Spencer.) Second, realism does not entail rationalism: the former is consistent with Meyerson's thesis about the unavoidable and irreducible irrational remainder left by every cognitive venture (Meyerson, 1908, pp. 272ff). Third, realism is not committed to the causal theory of perception, i.e., the hypothesis that sense impressions are causally related to physical objects. It is possible to hold to realism while assuming that perceptions, though caused by external (and internal) objects, are stochastically (rather than causally) related to them. In any event it is for science, not for epistemology, to find out what the mechanisms of perception and ideation are. Epistemology is concerned with knowledge (the end product) rather than with cognition (the process). Which brings us to our next theme.

3.4. *The Knowing Subject*

There is no knowledge without both an object of knowledge and a knowing subject. The claim that there is absolute knowledge, or knowledge in itself, above and beyond concrete knowing subjects, is fantastic. Moreover it violates the very syntax of 'to know', for 'x is known' is short for 'There is at least one y such that y is a knowing subject and y knows x'. Obliterate mankind and no human knowledge will remain. Furthermore every individual learns, imagines, and remembers (in sum, knows) in his own way: real cognition is as personal as ignorance. For this reason knowing or cognition is a subject for psychology. Epistemology takes knowing for granted and focuses on what is alleged to be known. In other words, epistemology is not concerned with personal knowledge, the only cognition there is. Epistemology makes the useful pretense that there exists *impersonal knowledge*, just as semantics pretends that there are propositions – not just judgments and sentences – and just as mathematics pretends that there are proofs whether or not they have actually been produced by anyone. To call what is known, i.e., knowledge, a *world* and assume that it is superimposed on the world of fact (Popper, 1968) is an unnecessary Platonic fantasy. There is only one world and cognitive subjects are part of it and intent on knowing (or ignoring) some chunks of it. Human action, whether cognitive or of another kind, does not create new worlds (except metaphorically) but variously transforms the only world there is.

Some philosophers, following certain suggestions of Peirce's and Morris', hold that personal knowledge is to be studied by a special discipline concerned also with other facets of human activity. This discipline, *pragmatics*, would consist of an empirical branch and a philosophical one. Philosophical (or pure) pragmatics would seek to establish logical relations among pragmatic concepts such as those of belief and intention (Martin, 1959) or to develop theories such as tense logic and the logic of personal pronouns, in which the subject or user plays a central role (Montague, 1968, 1970).

It may be questioned whether pure pragmatics, as currently developed, is on a sound methodological basis. Indeed if a discipline is concerned with factual items, such as users and circumstances of use, without benefiting from any empirical investigations, then it belongs with speculative metaphysics – with *Naturphilosophie* or with *Kulturphilosophie*. Theoretical pragmatics, if descriptive of fact, should be methodologically similar to theoretical psychology: it ought to build mathematical models of certain aspects of human behavior and have them tested in the laboratory. Only when normative might pragmatics be exempted from this requirement. For example, although actual problem solving is studied by psychology, the analysis of *well* conceived and *well* posed conceptual problems in general can be the concern of philosophers. That is, the logic and semantics of problems (like the logic and semantics of norms) belong to philosophical pragmatics. In conclusion, we have the following branching:

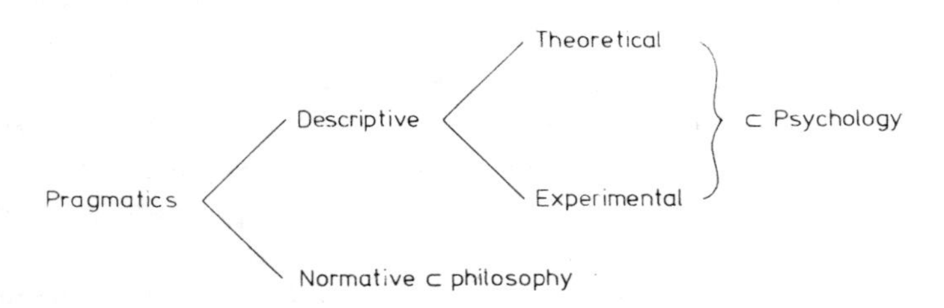

The relation between philosophical pragmatics and semantics could be this: each should investigate its own side of the coin of knowledge, but pragmatics would, in addition, take up problems concerning action. To put it differently: Epistemology can be split into two branches, according as the knowing subject is or is not reckoned with explicitly: (*a*) the study of personal knowledge, which is a branch of pragmatics, and (*b*) the study of impersonal knowledge, which overlaps with semantics. For example, the very concept of knowledge will split up into those of a set of statements, and opinion (or belief); the concept of truth into those of objective truth and personal truth; and the concept of meaning into those of semantic meaning and pragmatic meaning (or normal linguistic usage in a community).

It is well known that pragmatics is not married to pragmatism: that one may hold either a pragmatist view of pragmatics or an alternative conception of it. The extreme pragmatist theses about pragmatics seem to be these: (*a*) All genuine syntactical and semantical concepts have pragmatic counterparts, and (*b*) the former should all be reducible to (e.g., definable in terms of) pragmatic terms. While the first thesis is interesting, possibly true, and therefore deserves being investigated, the reduction thesis (*b*) is false. For example, Wittgenstein's reduction of meaning to use is, to put it charitably, a proposal to redefine 'meaning' so as to suit the purposes of the lexicographer. Usage is nothing but an (unreliable) indicator of significance. Besides, there is such a thing as correct (as distinct from widespread) use. And the relation of this normative concept to the semantic concept of significance is roughly this: The correct use of a sign is determined by its significance as revealed by an analysis of the system in which it occurs.

Moreover the pragmatic concepts presuppose syntactic and semantic concepts not the other way around. For example, the pragmatic statements $\ulcorner t$ has been proved from $A\urcorner$ and $\ulcorner x$ claims to have proved t from $A\urcorner$ presuppose the metamathematical concept of proof. Without the latter the previous statements would be unintelligible and untestable. Likewise and for the same reason the pragmatic proposition $\ulcorner$'a' means "b" to $c\urcorner$ presupposes a semantic concept of meaning. The same holds for belief assertions: In order to believe p we must first know p, which in turn presupposes p (quite apart from any truth value assignment). On the other hand we can pretend that there are propositions known to nobody. (This is what we do when we play around with the totality $\mathscr{C}n(A)$ of consequences of a given set A of assumptions.) And, of course, we can know p without believing p, just as we can believe p without professing to understand p.

A final warning: by an egocentric habit acquired in infancy we tend to employ pragmatic terms where they do not belong, thus contributing to an inflation of pragmatics. For example, we are prone to say that organic compounds are *found* in certain stars instead of asserting that *there are* organic compounds in stars; or that p is a possible result of *measuring* P rather than stating that p is a possible value of P; or that t

can be proved from A, instead of saying that t follows from A. The fact of the matter is that x may be found (or measured) if it *exists* to begin with, and that t may be proved provided it *is* a consequence. The phony use of pragmatic terms only creates confusion and fosters the illusion that pragmatics is a going concern rather than a project. In any case the semanticist's mode of speech is objective (subject-free) and tenseless: he considers matters *sub specie aeternitatis.*

So much for the relation of semantics to epistemology. Let us finally take a look at the most substantial yet also least reputable neighbor of semantics – metaphysics.

4. Metaphysics

4.1. *The Metaphysical Neutrality of Language*

We met metaphysics at the outset (Ch. 1, Sec. 3) when listing the kinds of object we should distinguish, as well as when discussing the nature of constructs. We also encountered metaphysics along the way, particularly when discussing reference, the representational function of some scientific constructs and, tangentially, factual truth. There are several other contacts between semantics and metaphysics. And, if the two are approached with a modicum of reason, their contacts need not be points of friction. Semantics can supply metaphysics with some tools for dissipating conceptual obscurity and confusion, and metaphysics can reciprocate by helping semanticists to avoid the extremes of vulgar materialism (e.g., literalism) and Platonism. We shall not pursue this line here. Instead, we will point to a place or two where semantics does *not* meet metaphysics. In fact we propose to discuss the alleged ontological commitment of language (the Whorf-Sapir doctrine) and, in particular, the claim that all existence sentences make such a commitment (Quine's thesis).

The celebrated Whorf-Sapir hypotheses are, in a nutshell, that (*a*) ordinary languages are loaded with world views or metaphysics, and (*b*) one's language determines at least in part the way he perceives and conceives the world. These two conjectures are, to say the least, gross exaggerations of the undeniable feedback of language into thought. It is false that every language reflects some world view, let alone some metaphysical system: any advanced language, by definition of 'advanced', can express any number of mutually incompatible world views, as witnessed by the diversity of metaphysics expounded in Sanskrit or in ancient Greek. And it is false that language is an important causal factor in perception and conception: a language permits or restricts our expressing what we perceive or conceive rather than what we perceive or conceive. Perception and conception occur in an epistemic framework not in a linguistic one: what determines in part whatever we perceive and conceive is experience, whether spontaneous or disciplined, personal or collective, rather than our mother tongue. Language is fortunately neutral with respect to our sensing and conceiving the world. Otherwise it would be impossible to formulate and discuss mutually incompatible views, and the very concept of expressive power of a language would be devoid of meaning.

Moreover the Whorf-Sapir doctrine is refuted by educational psychology: verbal teaching is inefficient unless the subject has already acquired some of the ideas he has been talked to about. If no words are available to express new ideas then one invents them at any age and in any culture. But if there are no ideas then words won't help – except to disguise the poverty in ideas, as Mephisto pointed out. Piaget and his coworkers have shown that a child who has learned the correct use of 'long', 'short' and cognate words may fail to understand that the amount of clay does not change when turned under his eyes from a ball into a sausage. Here again what matters is knowledge not language. In sum the Whorf-Sapir hypotheses are false: language is a tool that carries no ontological burden. It is rather the other way around: every ontology is constrained by the language it employs, in the sense that the latter may or may not be rich enough to express certain ontological ideas. (For example, ordinary language is insufficient to formulate a general theory of space.) Only theories can be ontically committed, and they are so provided they are not theories in logic or in pure mathematics. Language cannot even suggest a bad theory. But we went through this in Ch. 1: let us turn to a more interesting problem.

4.2. *The Metaphysical Neutrality of Logic*

If logic were nothing but a language then the arguments of the previous subsection would apply to it: logic should be ontologically neutral because it is the universal language. But logic is also and primarily a theory

– the theory of deduction – even though it can be used as a language (recall Ch. 1, Sec. 1.3). Therefore the relation of logic to metaphysics must be investigated separately. However, this is not the place for a thorough investigation of this problem, treated elsewhere (Bunge, 1974a). We must restrict our attention to those aspects of the question that involve semantics. We will limit ourselves to the question whether logic, in addition to being a reasoning organon, has a semantics similar to that of factual science. In other words, we shall ask the question whether logic involves a reference to reality or is ontologically neutral.

The thesis that logic does call for an ontological interpretation has been championed in recent times by Scholz and by Quine though on different grounds. According to Scholz (1941) the logical formulas hold for all possible worlds, ergo in particular for ours: logic would thus constitute a minimal ontology. This argument hinges on the identification of "possible world" and "model". But this is just a verbal trick: whereas a model is a conceptual object, the actual world is not a model of an abstract structure but the maximal self existent thing, and a possible world is something that neither logic nor ontology are in a position to characterize. Logic does not hold in all possible worlds – except metaphorically, in the sense that it is independent of the constitution and structure of the world. What model theory says is something more modest, namely, that a formula is a logical formula just in case it is satisfied in all models, i.e., if it holds under all interpretations of the nonlogical variables it contains. And the model independence of a logical identity consists in a particular matching of it with a whole set of constructs: it is an intraconceptual affair that has nothing to do with the actual world (Ch. 6, Sec. 2). Therefore the notion of universal validity (or analyticity) does not involve that of factual truth, which characterizes the semantics of factual science (Ch. 8). In conclusion, model theory is not concerned with any worlds and it does not exhibit logic as an ontology.

Unlike Scholz' approach, Quine's is direct: the latter finds ontology in the very bones of logic, particularly in quantification. Indeed he holds that "referential quantification is the key idiom of ontology" (Quine, 1969, p. 66). More explicitly: "Existence is what existential quantification expresses. There are things of kind F if and only if $(\exists x) Fx$. This is as unhelpful as it is undebatable, since it is how one explains the symbolic notation of quantification to begin with" (*op. cit.*, p. 97). The set of "things" a theory takes to exist is called by Quine the "ontological commitment", at times also the "ontology", of the theory (*op. cit.*, p. 106). Actually the word 'commitment' is too strong, since "that commitment can be momentary" (Quine, 1970, p. 99). That is, we have, if at all, ontological hypotheses.

If 'commitment' is unfortunate so are 'ontological' and 'ontic' in this context. Indeed, all that is intended to point to is the collection of *referents*, whether hypothesized or certified, intended or unintended, physical or conceptual, of a theory couched in some system of predicate logic. For example the referents of arithmetic are integers or perhaps the field of rationals. But a statement such as ⌜There are infinitely many prime numbers⌝ commits no one to believing in the *autonomous* existence of prime numbers. Far from being "an ontic idiom *par excellence*" (Quine, 1970, p. 92) existential quantification, unless qualified, is ontologically neutral. It is up to factual science, not to philosophy, to decide whether an unqualified existence statement has an ontic import.

The existential quantifier can be eliminated in favor of negation and the universal quantifier: "There are F's" is short for "It is not the case that everything fails to be an F". On the other hand, the concept of physical (real, ontic) existence cannot be so eliminated. In fact, to indicate physical existence we must assert unqualified existence and add that the objects in question happen to be physical objects – where the predicate 'is a physical object' is elucidated in ontology not in logic. There are *real* things of kind F if and only if $(\exists x)$ (Fx & x is a physical object). Only objects of this kind are of possible interest to ontology conceived, not as the general theory of objects of any kind, but as general cosmology. And only such statements of physical (real, ontic) existence will "commit" us to the objects they refer to – or rather to the task of finding out whether such objects are part of the actual world. (Incidentally, the collection of such referents of a statement constitutes its reference class, not its ontology: an ontology is a theory.)

Conceptual existence is parallel. There are *conceptual* objects of kind F if and only if $(\exists x)$ (Fx & x is a construct). This is the kind of existence the mathematician is interested in. Thus, notwithstanding the nominalist prohibition against classes, the mathematician deals with them all the time – only, he does not claim more than conceptual existence for them.

He has no reasons for thinking that the members of a set in a mathematical theory are more real than the set itself: if assumed or proved to be nonempty, a class enjoys the same conceptual existence as its members. Surely the metaphysician is entitled to asserting that classes have no physical existence – but then the mathematical individuals are just as unreal as classes, so that the metaphysician's remark is irrelevant.

In sum, while logic needs only a general concept of existence, the moment we cross the border of logic we need two specific concepts of existence – conceptual and physical. We have thus at least three different concepts of existence: neutral, conceptual, and physical. (These are concepts of existence not modes of being. Consequently there is no question of a general ontology branching off into an ontology of things and another of constructs. Let us not get carried away by words.) Let us emphasize the differences:

$(\exists x)\, Fx$	*Neutral existence*
$(\exists x)\, (Gx\ \&\ x$ is a construct$)$	*Conceptual existence*
$(\exists x)\, (Hx\ \&\ x$ is a physical object$)$	*Physical existence.*

Correspondingly we have three kinds of existence problem: neutral, conceptual, and physical. The last two kinds are irreducible. Thus the question of the (conceptual) existence of classes has nothing to do with the question of the (physical) existence of things: this is why conceptualism is compatible with ontological individualism, or the belief that only individuals can have physical existence. And, since all classes are constructs, infinite sets pose no special problem of existence.

Nor are those kinds of existence question reducible to a fourth, more basic type of question. In particular it is not possible to reduce all questions of existence to the existence *in* or *of* a linguistic framework – the so called internal and external questions (Carnap, 1950). Thus the problem whether a "theoretical entity", such as the parton or the black hole, has a real counterpart is not a question concerning exclusively a man made "framework", let alone a linguistic one: it is an empirical problem – namely that of subjecting the factual theory concerned to observational tests. To hold that physical existence questions are not scientific but concern linguistic frameworks is no better than claiming, with Mach and the Vienna Circle, that those are sheer metaphysical questions. In either case the typical features of the semantics of scientific theories are missed. (For further criticisms see Ferrater Mora (1967).)

The upshot of our discussion is this: (*a*) just as we distinguish constructs from things so we must distinguish conceptual existence from physical existence; (*b*) logic employs a generic or neutral concept of existence, hence it is not in a position to assert or deny anything concerning the real world: it is ontologically neutral; (*c*) mathematics uses the more special notion of conceptual existence but it is equally unconcerned with reality; (*d*) factual science and metaphysics utilize the concept of physical (material) existence.

4.3. *Metaphysical Commitments of the Semantics of Science*

Logic, mathematics, and their semantics (i.e., model theory) are metaphysically neutral: they do not even have to assume the existence of any objects other than constructs and their symbols. Not so the semantics of factual science. Indeed the latter has the following presuppositions of a metaphysical character: (i) there are both constructs (in particular concepts) and physical objects (in particular signs); (ii) some signs designate constructs and some constructs refer to physical objects; (iii) the theories referring to physical objects constitute more or less adequate (true) representations of aspects of the world.

If the above assumptions were denied then the very concepts of designation, denotation, reference, representation, and factual truth would become pointless. Our own realist brand of semantics adds two further assumptions: (iv) constructs are ideal objects: unlike physical objects, constructs have no separate (physical) existence, and (v) the putative referents of a factual construct are supposed (rightly or wrongly) to be self existent not just possibilities of perception.

This is all the metaphysical load our semantics carries. It is necessary and sufficient to make sense of the rules of denotation and of the semantic assumptions included in scientific theories (when reconstructed according to our semantics) as well as to make sense of the testing for truth of fact and of the correction of hypotheses and theories in the light of fresh experience. Beyond this point there is room for any number of possible ontological theories sketching the basic constitution and structure of the world. The author's own preference lies with cross disciplinary metaphysical theories, mathematical in form and contiguous with factual science (Bunge, 1971b). But this is another story to be told in a different work. (Cf. Volumes 3 and 4 of this *Treatise*.)

5. Parting words

Up till now philosophical semantics has led a rather secluded life: it has had hardly any contacts with other branches of philosophy and with factual science. Whence its main failing: its nearly total irrelevance to anything lying outside logic and mathematics – and consequently its inability to be of any help in understanding what goes on in that external world.

It has become an urgent task to reset semantics in its philosophic context – the one in which it used to dwell from Socrates through Buridan to Russell – as well as to open it up to the winds of factual science, without thereby relinquishing the ideals of rigor and systemicity bequeathed upon us by Tarski and Carnap.

Secondly, semantics should become a service discipline like logic and mathematics – one ever ready to offer a helping hand to whichever neighbor may need it. (See Table 10.1.) But in order to become useful semantics must mix with its neighbors and even meddle with their affairs, thus learning their ways and needs. The watchword should no longer be 'Come and behold my neatness' but rather 'Let us go and see how we can fix that mess'.

Thirdly, semantics should be freed from its associations with obsolete philosophies, and fresh semantic theories answering the needs of science and inspired in critical realism should be built. We need far more work in realist semantics and we should look everywhere for inspiration and control: a philosophy that minds only its own business does not mind its main business.

May this work contribute to implementing Kanenas T. Pota's ideal of "an exact, systematic, realist, and above all useful semantics". In any case this semantics of ours is to serve as a prolegomenon to the substantive portions of the philosophical system set forth in this *Treatise*, namely ontology, epistemology, and axiology.

TABLE 10.1
Semantics and its neighbors

Trait	Science	Mathematics	Philosophy			
			Logic	Semantics	Epistemology	Metaphysics
Referents	Things (concrete systems)	Conceptual structures (categories)	Constructs	Constructs & signs	Knowledge	Reality
Range	Aspects and levels of things	Theories	Universal	Universal	Universal	Universal
Truth type	Factual	Formal	Logical	Philosophical	Philosophical	Philosophical
Accuracy	Limited	Complete (except in numerical analysis)	Complete	Complete with reference to formal science	Limited	Limited
Method	Hypothesis & theory, observation & experiment	Postulation & proof, example & counter-example	Postulation & proof, example & counter-example	Postulation, proof, & checking with logic, mathematics, or science	Analysis, postulation, & checking with substantive knowledge & methodology	Postulation & checking with science
Goals	Finding laws, describing, explaning & predicting	Building theories about conceptual structures & inter-relating them	Depuration & systematization of general methods of analysis and proof	Elucidation & articulation of concepts of meaning, truth & cognates	Elucidation & articulation of all the concepts about factual knowledge & ignorance	Finding overall (cross-disciplinary) structures & patterns in the world

Table 10.1 (Continued)

Trait	Science	Mathematics	Philosophy			
			Logic	Semantics	Epistemology	Metaphysics
Roles	Basing technology, controlling world views	Forging conceptual tools for science & philosophy	Keeping watch on reasoning	Conceptual hygiene, spotting genuine referents, clarifying sense, and scotching myths in the philosophy of scientists	Methodological alertness & open-mindedness	Elucidating concepts common to several sciences, posing fruitful problems & proposing illuminating hypotheses about the world
Employs	Logic, mathematics & philosophy	Logic	Mathematics	Logic, mathematics & science	Logic, mathematics & science	Logic, mathematics & science

BIBLIOGRAPHY

Addison, J. W., L. Henkin, and A. Tarski, Eds. (1965), *The Theory of Models*. Amsterdam: North-Holland Publ. Co.

Ajdukiewicz, K. (1967a). Intensional expressions. *Studia Logica* **20**: 63–86.

Ajdukiewicz, K. (1967b). Proposition as the connotation of sentence. *Studia Logica* **20**: 87–98.

Alston, W. P. (1968). Meaning and use. In Parkinson Ed. 1968: 141–165.

Arnauld, A. and P. Nicole (1662). *La logique, ou l'art de penser*. Paris: Flammarion 1970.

Attfield, R. and M. Durrant (1973). The irreducibility of 'meaning'. *Nous* **VII**: 282–298.

Austin, J. L. (1962). *How to do Things with Words*. London: Oxford University Press.

Ayer, J. A. (1959). *Logical Positivism*. Glencoe, Ill.: Free Press.

Bar-Hillel, Y. (1970). *Aspects of Language*. Jerusalem: Magnes Press.

Bar-Hillel, Y., E. I. J. Poznanski, M. O. Rabin, and A. Robinson, Eds. (1961). *Essays on the Foundations of Mathematics*. Jerusalem: Magnes Press.

Bar-Hillel, Y., Ed. (1965). *Logic, Methodology and Philosophy of Science*. Amsterdam: North-Holland Publ. Co.

Barcan-Marcus, R. (1962). Interpreting quantification. *Inquiry* **5**: 252–259.

Bell, J. L. and A. B. Slomson (1969). *Models and Ultraproducts*. Amsterdam: North-Holland.

Benacerraf, P. (1973). Mathematical truth. *Journal of Philosophy* **70**: 661–679.

Beth, E. W. (1962). Extension and intension. In *Logic and Language: Studies dedicated to Professor Rudolf Carnap*. Dordrecht: D. Reidel Publ. Co.

Birkhoff, G. (1961). *Lattice Theory*, 2nd ed. Providence, R. I.: American Mathematical Society.

Black, M. (1962). *Models and Metaphors*. Ithaca: Cornell University Press.

Blanshard, B. (1939). *The Nature of Thought*. London: Allen & Unwin; N. York: Mcmillan.

Blumenthal, L. M. and K. Menger (1970). *Studies in Geometry*. San Francisco: W. H. Freeman and Co.

Bolzano, B. (1837). *Wissenschaftslehre*, 4 vols. Sulzbach: Seidelsche Buchhandlung. Abridged Engl. transl. by R. George, *Theory of Science*. Berkeley and Los Angeles: University of California Press 1972.

Bolzano, B. (1851). *Paradoxes of the Infinite*. Transl. D. A, Steele. London: Routledge & Kegan Paul, 1950.

Bunge, M. (1957). Lagrangian formulation and mechanical interpretation. *Amer. J. Phys.* **25**: 211–218.

Bunge, M. (1959a). *Causality*, Cambridge, Mass.: Harvard University Press. Repr.: The World Publ. Co. 1963.

Bunge, M. (1959b). *Metascientific Queries*. Springfield, Ill.: Charles C. Thomas Publ.

Bunge, M. (1961a). Laws of physical laws. *Amer. J. Phys.* **29**: 518–529. Repr. in Bunge (1963a).

Bunge, M. (1961b). Analyticity redefined. *Mind* **70**: 239–246.

Bunge, M. (1961c). The weight of simplicity. *Phil. Sci.* **28**: 120–149.

Bunge, M. (1962). *Intuition and Science*. Englewood Cliffs, N. J.: Prentice Hall.
Bunge, M. (1963a). *The Myth of Simplicity*. Englewood Cliffs, N.J.: Prentice-Hall.
Bunge, M. (1963b). A general black box theory. *Phil. Sci.* **30**: 346–358.
Bunge, M. (1964). Phenomenological theories. In M. Bunge, Ed., *The Critical Approach*, pp. 234–254. New York: Free Press.
Bunge, M. (1966). Mach's critique of Newtonian mechanics. *Amer. J. Phys.* **34**: 585–596.
Bunge, M. (1967a). *Scientific Research*, 2 vols. New York: Springer-Verlag.
Bunge, M. (1967b). *Foundations of Physics*. New York: Springer-Verlag.
Bunge, M. (1967c). Analogy in quantum mechanics: from insight to nonsense. *Brit. J. Phil. Sci.* **18**: 265–286.
Bunge, M. (1967d). Physical axiomatics. *Rev. Mod. Phys.* **39**: 463–474.
Bunge, M. (1967e). A ghost-free axiomatization of quantum mechanics. In M. Bunge, Ed., *Quantum Theory and Reality*, pp. 98–117. New York: Springer-Verlag.
Bunge, M. (1969). What are physical theories about? *Amer. Philos. Quart. Monograph* **3**: 61–99.
Bunge, M. (1970a). Theory meets experience. In H. Kiefer and M. Munitz, Eds., *Contemporary Philosophic Thought*, **2**: 138–165. Albany, N.Y.: State University of New York Press.
Bunge, M. (1970b). Problems concerning inter-theory relations. In P. Weingartner and G. Zecha, Eds., *Induction, Physics, and Ethics*, pp. 287–325. Dordrecht: D. Reidel Publ. Co.
Bunge, M. (1971a). A mathematical theory of the dimensions and units of physical quantities. In M. Bunge, Ed., *Problems in the Foundations of Physics*, pp. 1–16. New York: Springer-Verlag.
Bunge, M. (1971b). Is scientific metaphysics possible? *J. Phil.* **68**: 507–520.
Bunge, M. (1971c). A new look at definite descriptions. *Phil. Sci. (Japan)* **4**: 131–146.
Bunge, M. (1972a). A program for the semantics of science. *J. Phil. Logic* **1**: 317–328. Reprinted in Bunge, Ed. (1973a).
Bunge, M. (1972b). Metatheory. In *Scientific Thought*. Paris–The Hague: Mouton/Unesco.
Bunge, M. (1973a). *Method, Model and Matter*. Dordrecht: D. Reidel Publ. Co.
Bunge, M. (1973b). *Philosophy of Physics*. Dordrecht: D. Reidel Publ. Co.
Bunge, M. (1973d). Meaning in science. *Proc. XVth World Congress of Philosophy* **2**: 281–286. Sofia.
Bunge, M., Ed. (1973a). *Exact Philosophy: Problems, Goals, and Methods*. Dordrecht: D. Reidel Publ. Co.
Bunge, M. (1974a). The relations of logic and semantics to ontology. *J. Phil. Logic* **3**: 195–210.
Bunge, M. (1974b). Possibility and probability. In W. L. Harper and C. A. Hooker, Eds., *Proc. of an Internat. Research Colloquium on Foundations of Probability and Statistics*, Vol. 3. Dordrecht and Boston: D. Reidel Publ. Co.
Burton, W. K. (1968). Constructive thermodynamics. In A. A. Schmidt, K. Schütte, and H.-J. Thiele, Eds., *Contributions to Mathematical Logic*, pp. 75–89. Amsterdam: North-Holland.
Campbell, D. T. (1959). Methodological suggestions from a comparative psychology of knowledge processes. *Inquiry* **2**: 152–82.
Carathéodory, C. (1924). Zur Axiomatik der speziellen Relativitätstheorie. *Sitz. ber. preuss. Akad. Wiss. Physik.-math. Kl.*, **12**.
Carnap, R. (1928). *Scheinprobleme in der Philosophie*. Engl. transl. in *The Logical Structure of the World*. Berkeley and Los Angeles: University of California Press, 1967.

Carnap, R. (1936). Testability and meaning. *Phil. Sci.* **3**: 419–471; **4**: 1–40.
Carnap, R. (1939). *Foundations of Logic and Mathematics*. Chicago: University of Chicago Press.
Carnap, R. (1942). *Introduction to Semantics*. Cambridge, Mass.: Harvard University Press.
Carnap, R. (1947). *Meaning and Necessity*. Chicago: University of Chicago Press, Enlarged ed. 1956.
Carnap, R. (1950). Empiricism, semantics, and ontology. *Rev. Intern. de Philos.* **4**: 20–40. Repr. in Linsky, Ed. (1952).
Carnap, R. (1952). Meaning postulates. *Phil. Studies* **3**: 65–73.
Carnap, R. (1956). The methodological character of theoretical concepts. In H. Feigl and M. Scriven, Eds., *Minnesota Studies in the Philosophy of Science*, I: 38–76. Minneapolis: University of Minnesota Press.
Carnap, R. (1958). *Introduction to Symbolic Logic and its Applications*. New York: Dover Publications, Inc.
Carnap, R. (1961). On the use of Hilbert's ε-operator in scientific theories. In Bar-Hillel *et al.*, Eds. (1966), 156–164.
Carnap, R. (1963a). Intellectual autobiography. In Schilpp Ed. (1963), pp. 3–84.
Carnap, R. (1963b). Replies and systematic expositions. In Schilpp Ed. (1963), pp. 859–1013.
Castonguay, C. (1972). *Meaning and Existence in Mathematics*. Wien-New York: Springer-Verlag.
Chang, C. C. and H. J. Keisler (1966). *Continuous Model Theory*. Princeton, N.J.: Princeton University Press.
Chang, C. C. and H. J. Keisler (1973). *Model Theory*. Amsterdam-London: North-Holland Publ. Co.
Church, A. (1951). A formulation of the logic of sense and denotation. In P. Henle, H. M. Kallen and S. K. Langer, Eds., *Structure, Method and Meaning: Essays in Honor of Henry M. Scheffer*, pp. 3–24. New York: The liberal Arts Press.
Church, A. (1973/74). Outline of a revised formulation of the logic of sense and denotation. *Nous* **7**: 24–33, **8**: 135–156.
Cole, M. and I. Maltzman, Eds. (1969). *A Handbook of Contemporary Soviet Psychology*. New York and London: Basic Books, Inc.
Davidson, D. (1967). Truth and meaning. *Synthese* **17**: 304–333. Repr. in J. W. Davis, D. J. Hockney, and W. K. Wilson, Eds., *Philosophical Logic*. Dordrecht: D. Reidel Publ. Co., 1969.
DeWitt, B. S. (1970). Quantum mechanics and reality. *Phys. Today* **23**, No. 9, 30–35.
Dingler, H. (1907). *Grundlinien einer Kritik und exakten Theorie der Wissenschaften, insbesondere der Mathematik*. München: Ackermann.
Dummett, M. (1973). *Frege. Philosophy of Language*. London: Duckworth & Co. Ltd.
Dirac, P. A. M. (1942). The physical interpretation of quantum mechanics. *Proc. Roy. Soc. (A)* **180**: 1–40.
Eberle, R. A. (1969). Denotationless terms and predicates expressive of positive qualities. *Theoria* **35**: 104–123.
Einstein, A. (1936). Physics and reality. *J. Franklin Institute* **221**: 313–347.
Enriques, F. (1943). *Problems of Science*. La Salle, Ill.: Open Court.
Everett III, H. (1957). "Relative state" formulation of quantum mechanics. *Rev. Mod. Phys.* **29**: 454–462.
Feigl, H. (1958). The 'mental' and the 'physical'. In H. Feigl, M. Scriven, and G. Maxwell, Eds., *Minnesota Studies in the Philosophy of Science*, II: 370–497.

Feller, W. (1968). *An Introduction to Probability Theories and its Applications* I, 3rd ed. New York and London: John Wiley & Sons.
Ferrater Mora, J. (1967). *El ser y el sentido.* Madrid: Revista de Occidente.
Feyerabend, P. K. (1970). Against method: Outline of an anarchistic theory of knowledge. In Radner and Winokur Eds. (1970) pp. 17–130.
Fine, A. (1968). Logic, probability, and quantum mechanics. *Phil. Sci.* **35**: 101–111.
Fréchet, M. (1939). The diverse definitions of probability. *Erkenntnis* **8**: 7–23.
Frege, G. (1879). Begriffsschrift. Repr. in *Begriffsschrift und andere Aufsätze*. I. Angelelli, Ed. Hildesheim: Georg Olms 1964.
Frege, G. (1891). Funktion und Begriff. In Angelelli (1967).
Frege, G. (1892). Über Sinn und Bedeutung. In Angelelli (1967).
Frege, G. (1894). Rezension von: E. G. Husserl, Philosophie der Arithmetik. I. In Angelelli (1967).
Frege, G. (1912). Anmerkungen zu P. E. B. Jourdain, The development of the theories of mathematical logic, etc. In Angelelli (1967).
Frege, G. (1969). *Nachgelassene Schriften.* H. Hermes, F. Kambartel and F. Kaulbach, Eds. Hamburg: F. Meiner.
Freudenthal, H. (1970). What about the foundations of physics? *Synthese* **21**: 93–106.
Freudenthal, H. (1971). More about *Foundations of Physics*. *Foundations of Physics* **1**: 315–323.
Gentilhomme, Y. (1968). Les ensembles flous en linguistique. *Cahiers de linguistique théorique et appliquée* **5**: 47–65.
Giles, R. (1964). *Mathematical Foundations of Thermodynamics*. London: Pergamon.
Goguen, J. A. (1969). The logic of inexact concepts. *Synthese* **19**: 325–373.
Conseth, F. (1938). *La méthode axiomatique*. Paris: Gauthier-Villars.
Goodman, N. (1960). The way the world is. *Rev. Metaphys.* **14**: 48–56.
Goodman, N. (1968). *The Languages of Art*. Indiana and New York: Bobbs Merrill Co.
Goodstein, R. L. (1968). Existence in mathematics. *Compositio mathematica* **20**: 70–82.
Goodstein, R. L. (1969). Empiricism in mathematics. *Dialectica* **23**: 50–57.
Grossberg, S. (1969). Some networks that can learn, remember, and reproduce. *I. J. Math. and Mechanics* **19**: 53–91.
Harrison, M. E. (1965). *Introduction to Switching and Automata Theory*. New York: McGraw-Hill Book Co.
Hartnett, W. E. (1963). *Principles of Modern Mathematics*, 1. Chicago: Scott, Foresman, and Co.
Hartnett, W. E. (1970). *Principles of Modern Mathematics*, 2. Chicago: Scott, Foresman, and Co.
Heisenberg, W. (1955). The development of the interpretation of the quantum theory. In W. Pauli, L. Rosenfeld, and V. Weisskopf, Eds., *Niels Bohr and the Development of Physics*, pp. 12–29. London: Pergamon.
Helmholtz, H. v. (1873). *Lectures on Scientific Subjects*. London: Longmans, Green and Co.
Hempel, C. G. (1970). On the "standard conception" of scientific theories. In M. Radner and S. Winokur, Eds., (1970) pp. 142–163.
Henkin, L., P. Suppes and A. Tarski, Eds. (1959). *The Axiomatic Method*. Amsterdam: North-Holland Publ. Co.
Hermes, H. (1963). *Einführung in die mathematische Logik*. Stuttgart: B. G. Teubner.
Hesse, M. (1965). The explanatory function of metaphor. In Bar-Hillel, ed. (1965), pp. 249–259.
Hilbert, D. (1924). Grundlagen der Physik. In *Gesammelte Abhandlungen*, vol. 3. Berlin:

Julius Springer, 1935.
Hilbert, D. and P. Bernays (1968). *Grundlagen der Mathematik* I, 2nd ed. Berlin-Heidelberg-New York: Springer-Verlag.
Hintikka, J. (1962). *Knowledge and Belief.* Ithaca, N.Y.: Cornell University Press.
Hintikka, J. (1969). *Models for Modalities.* Dordrecht: D. Reidel Publ. Co.
Hooker, C. A. (1974). Systematic realism. *Synthese* **26**: 409–497.
Jost, R. (1965). *The General Theory of Quantized Fields.* Providence, R. I.: Amer. Mathematical Society.
Kalish, D. and R. Montague (1957). Remarks on descriptions and natural deduction. *Arch. f. Math. Logik u. Grundlagenforschung* **3**: 50–73.
Kalmár, L. (1967). Foundations of mathematics – whither now? In I. Lakatos, Ed., *Problems in the Philosophy of Mathematics*, pp. 187–194. Amsterdam: North-Holland Publ. Co.
Katz, J. J. and J. A. Fodor (1963). The structure of a semantic theory. *Language* **39**: 170–210.
Kemeny, J. G. (1956). A new approach to semantics. *J. Symbol. Logic* **21**: 1–27, 149–161.
Kleiner, S. A. (1971). Criteria for meaning changes. In M. Bunge, Ed., *Problems in the Foundations of Physics*, pp. 131–144. New York: Springer-Verlag.
Klemke, E. D., Ed. (1968). *Essays on Frege.* Urbana, Ill.: University of Illinois Press.
Kolmogoroff, A. N. (1963). The theory of probability. In A. D. Aleksandrov, A. N. Kolmogoroff, and M. A. Lavrent'ev, Eds., *Mathematics: Its Content, Methods, and Meaning*, 3 vols. Cambridge, Mass.: M.I.T. Press.
Körner, S. (1964). Deductive unification and idealisation. *Brit. J. Phil. Sci.* **14**: 274–284.
Kripke, S. (1959). A completeness theorem in modal logic. *Journal of Symbolic Logic* **24**: 1–15.
Kuhn, T. (1962). *The Structure of Scientific Revolutions.* Chicago: University of Chicago Press.
Landau, L. D. and E. M. Lifshitz (1958). *Quantum Mechanics.* London: Pergamon Press.
Leibniz, G. W. v. (1703). *Nouveaux essais sur l'entendement humain.* Paris: Flammarion, s.d.
Leibniz, G. W. v. (1714). The Monadology. In *Philosophical Papers and Letters*, Ed. L. E. Loemker, vol. 2. Chicago: University of Chicago Press 1956.
Lenin, V. I. (1909). *Materialism and Empirio-Criticism.* Moscow: Foreign Language Publ. House 1947.
Lewis, C. I. (1929). *Mind and the World Order.* New York: Scribner.
Lewis, C. I. (1944). The modes of meaning. *Philosophy and Phenomenological Research* **4**: 236–249.
Lewis, C. I. (1951). Notes on the logic of intensions. In P. Henle *et al.*, Eds., *Structure, Method and Meaning.* New York: Liberal Arts Press.
Linsky, L., Ed. (1952). *Semantics and the Philosophy of Language.* Urbana: University of Illinois Press.
Lorenzen, P. (1967). Moralische Argumentationen im Grundlagenstreit der Mathematik. In *Tradition und Kritik: Festschrift für R. Zocher*, pp. 219–227. Stuttgart: F. Fromman.
Luria, A. R. (1969). Speech development and the formation of mental processes. In Cole and Maltzmann, Eds. 1969, pp. 121–162.
Łukasiewicz, J. (1913). Logical foundations of probability theory. In *Selected Works*, Ed. L. Borkowski. Amsterdam: North-Holland Publ. Co. 1970.
MacLane, S. and G. Birkhoff (1967). *Algebra.* New York: Macmillan.
Martin, R. M. (1958). *Truth and Denotation.* Chicago: University of Chicago Press.
Martin, R. M. (1959). *Toward a Systematic Pragmatics.* Amsterdam: North-Holland Publ. Co.

Maxwell, J. C. (1871). Remarks on the mathematical classification of physical quantities. *Proc. London Math. Soc.* **3**: 224–233.
Mendelson, E. (1963). *Introduction to Mathematical Logic*. Princeton, N.J.: D. Van Nostrand Co., Inc.
Meyerson, E. (1908). *Identité et réalité*. Paris: Félix Alcan.
Mill, J. S. (1875). *A System of Logic*, 8th ed. London: Longmans, Green and Co. 1952.
Moisil, Gr. C. (1972). *Essais sur les logiques non chrysippiennes*. Bucarest: Editions de l'Académie de la R. S. de Roumanie.
Montague, R. (1968). Pragmatics. In R. Klibansky, Ed., *Contemporary Philosophy* I: 102–122. Firenze: La Nuova Italia Editrice.
Montague, R. (1970). Pragmatics and intensional logic. *Synthese* **22**: 68–94.
Naess, A. (1956). Synonymity and empirical research. *Methodos* **8**: 3–22.
Nagel, E. (1956). *Logic Without Metaphysics*. Glencoe, Ill.: The Free Press.
Nagel, E., P. Suppes, and A. Tarski, Eds. (1962). *Logic, Methodology and Philosophy of Science*. Stanford: Stanford University Press.
Padoa, A. (1901). Essai d'une théorie algébrique des nombres entiers, précédé d'une introduction à une théorie déductive quelconque. *Bibliothèque du Congrès International de Philosophie* **3**: 309–365.
Parkinson, J. H. R. (1968). *The Theory of Meaning*. Oxford: Oxford University Press.
Peano, G. (1921). Le definizioni in matematica. *Periodico di matematiche* **1**: 175–189.
Popper, K. R. (1959). The propensity interpretation of probability. *Brit. J. Phil. Sci.* **10**: 25–42.
Popper, K. R. (1963a). The demarcation between science and metaphysics. In Schilpp, Ed. (1963), pp. 183–226.
Popper, K. R. (1963b). *Conjectures and Refutations*. New York: Basic Books.
Popper, K. R. (1968). Epistemology without a knowing subject. In B. van Rootselaar and J. F. Staal, Eds. (1968), pp. 333–373.
Popper, K. R. (1970). A realist view of logic, physics, and history. In W. Yourgrau and A. D. Breck, Eds. (1970), pp. 1–37.
Popper, K. R. (1972). *Objective Knowledge*. Oxford: Clarendon Press.
Popper, K. R. (1974). Autobiography. In Schilpp, Ed. (1974).
Przelecki, M. (1969). *The Logic of Empirical Theories*. London: Routledge & Kegan Paul.
Putnam, H. (1962). What theories are not. In Nagel, Suppes and Tarski, Eds. (1962), pp. 240–251.
Quine, W. V. (1952). *From a Logical Point of View*. Cambridge, Mass.: Harvard University Press.
Quine, W. V. (1960). *Word and Object*. Cambridge, Mass.: M.I.T. Press.
Quine, W. V. (1964). Implicit definition sustained. *J. Phil.* **60**: 71–74.
Quine, W. V. (1969). *Ontological Relativity and Other Essays*. New York: Columbia University Press.
Quine, W. V. (1970a). Existence. In W. Yourgrau and A. D. Breck, Eds. (1970), pp. 89–98.
Quine, W. V. (1970b). *Philosophy of Logic*. Englewood Cliffs, N.J.: Prentice-Hall.
Quine, W. V. and N. Goodman (1940). Elimination of extra-logical postulates. *J. Symbol. Logic* **5**: 104–109.
Radner, M. and S. Winokur, Eds. (1970). *Minnesota Studies in the Philosophy of Science* IV. Minneapolis: University of Minnesota Press.
Ramsey, F. P. (1931). *The Foundations of Mathematics*. London: Routledge & Kegan Paul.
Rapoport, A. (1966). *Two-Person Game Theory*. Ann. Arbor: University of Michigan Press.
Reichenbach, H. (1944). *Philosophic Foundations of Quantum Mechanics*. Berkeley and Los

Angeles: University of California Press.

Reichenbach, H. (1949). *The Theory of Probability*. Berkeley and Los Angeles: University of California Press.

Rescher, N. (1973). *The Coherence Theory of Truth*. Oxford: Clarendon Press.

Robinson, A. (1963). *Introduction to Model Theory and to the Metamathematics of Algebra*. North-Holland Publ. Co.

Robinson, A. (1965). Formalism 64. In Bar-Hillel, Ed. (1965), pp. 228–246.

Rootselaar, B. van, and J. F. Staal, Eds. (1968). *Logic, Methodology and Philosophy of Science* III. Amsterdam: North-Holland Publ. Co.

Rosen, R. (1970). *Dynamical System Theory in Biology*. New York: Wiley-Interscience.

Rosenfeld, L. (1961). Foundations of quantum theory and complementarity. *Nature* **190**: 384–388.

Rozeboom, W. (1962). The Factual content of theoretical concepts. In H. Feigl and G. Maxwell, Eds., *Minnesota Studies in the Philosophy of Science* III: 273–357. Minneapolis, Min.: University of Minnesota Press.

Russell, B. (1905). On denoting. *Mind N.S.* **14**: 479–493.

Russell, B. (1919a). *Introduction to Mathematical Philosophy*. London: George Allen & Unwin.

Russell, B. (1919b). On propositions: what they are and how they mean. In R. C. Marsh, Ed., *Logic and Knowledge*. London: George Allen & Unwin 1956.

Salt, D. (1971). Physical axiomatics: Freudenthal vs. Bunge. *Foundations of Phys.* **1**: 307–313.

Scheibe, E. (1973). The approximate explanation and the development of physics. In P. Suppes, L. Henkin, A. Joja, Gr. C. Moisil, Eds., *Logic, Methodology and Philosophy of Science IV*, pp. 931–942. Amsterdam: North-Holland.

Schilpp, P. A., Ed. (1963). The Philosophy of Rudolf Carnap. La Salle, Ill.: Open Court: London: Cambridge University Press.

Schilpp, P. A., Ed. (1974). *The Philosophy of Karl R. Popper*. La Salle, Ill.: Open Court.

Schlick, M. (1932/33). Positivism and realism. In Ayer, Ed. 1959.

Scholz, H. (1941). *Metaphysik als strenge Wissenschaft*. Repr.: Darmstadt: Wissenschaftliche Buchgesellschaft 1965.

Scholz, H. (1969). *Mathesis Universalis*, 2nd ed. Darmstadt: Wissenschaftliche Buchgesellschaft.

Scott, D. (1967). Existence and description in formal logic. In R. Schoenman, Ed., *Philosopher of the Century: Essays in Honour of Bertrand Russell*, pp. 181–200. London: Allen & Unwin.

Scott, D. and C. Strachey (1971). *Towards a mathematical semantics for computer languages*. Oxford: Oxford University Computer Lab.

Shoenfield, J. R. (1967). *Mathematical Logic*. Reading, Mass.: Addison-Wesley.

Stenlund, S. (1973). *The Logic of Description and Existence*. Uppsala: Philosophical Studies.

Strawson, P. F. (1950). On referring. *Mind, N.S.* **59**: 320–344.

Strawson, P. F. (1964). Identifying reference and truth-values. *Theoria* **30**: 96–118.

Strawson, P. F. (1971). *Logico-Linguistic Papers*. London: Methuen & Co. Ltd.

Suppes, P. (1957). *Introduction to Logic*. Princeton, N.J.: D. Van Nostrand.

Suppes, P. (1959). Measurement, empirical meaningfulness, and three-valued logic. In C. West Churchman and P. Ratoosh, Eds., *Measurement: Definitions and Theories*. New York: Wiley.

Suppes, P. (1961). A comparison of the meaning and uses of models in mathematics and

the empirical sciences. In H. Freudenthal, Ed., *The Concept and the Role of the Model in Mathematics and Natural and Social Sciences.* Dordrecht: D. Reidel Publ. Co.
Suppes, P. (1965). Logics appropriate to empirical theories. In Addison *et al.*, Eds. (1965), pp. 364–375.
Suppes, P. (1967). *Set-theoretical Structures in Science.* Stanford University: Institute for Mathematical Studies in the Social Sciences.
Suppes, P. (1969). *Studies in the Methodology and Foundations of Science.* Dordrecht: D. Reidel.
Svenonius, L. (1973). Translation and reduction. In Bunge, Ed. (1973a).
Tarski, A. (1934). Some methodological investigations on the definability of concepts. In Tarski (1956).
Tarski, A. (1936). Der Wahrheitsbegriff in den formalisierten Sprachen. In Tarski 1956.
Tarski, A. (1954). Contributions to the theory of models I, II. *Indagationes Mathematicae* **16**: 572–588.
Tarski, A. (1955). Contributions to the theory of models III. *Indagationes Mathematicae* **17**: 56–64.
Tarski, A. (1956). *Logic, Semantics, Metamathematics.* Oxford: Clarendon Press.
Tarski, A., A. Mostowski, and R. M. Robinson (1953). *Undecidable Theories.* Amsterdam: North-Holland Publ. Co.
Truesdell, C. and R. Toupin (1960). The classical field theories. In S. Flügge, Ed., *Handbuch der Physik* III/1. Berlin-Göttingen-Heidelberg: Springer-Verlag.
Tuomela, R. (1973). *Theoretical Concepts.* Wien and New York: Springer-Verlag.
Waissman, F. (1955). Verifiability. In A. Flew, Ed., *Logic and Language.* Oxford: Basil Blackwell.
Wang, H. (1951). Arithmetic translation of axiom systems. *Trans. Amer. Math. Soc.* **71**: 283–293.
Wang, H. (1966). Process and existence in mathematics. In Bar-Hillel *et al.*, Eds. (1966), pp. 328–351.
Weyl, H. (1919). *Raum-Zeit-Materie*, 3rd ed. Berlin: J. Springer.
Wheeler, J. A. (1957). Assessment of Everett's "relative state" formulation of quantum theory. *Rev. Mod. Phys.* **29**: 463–465.
White, H. J. and S. Tauber (1970). *Systems Analysis.* Philadelphia, Pa.: W. B. Saunders.
Whitehead, A. N. (1898). *A Treatise on Universal Algebra* I. Cambridge: Cambridge University Press.
Whitehead, A. N. and B. Russell (1927). *Principia Mathematica* I, 2nd ed. Cambridge: Cambridge University Press.
Williams, D. C. (1937). The realistic interpretation of scientific sentences. *Erkenntnis* **7**: 169–178, 375–382.
Williams, D. C. (1966). *Principles of Empirical Realism.* Springfield, Ill.: Charles C. Thomas.
Winograd, T. (1972). *Understanding Natural Language.* New York: Academic Press.
Yourgrau, W. and A. D. Breck, Eds. (1970). *Physics, Logic, and History.* New York and London: Plenum Press.
Zinov'ev, A. A. (1973). *Foundations of the Logical Theory of Scientific Knowledge* (Complex Logic). Dordrecht: D. Reidel Publ. Co.

INDEX OF NAMES

INDEX OF SUBJECTS

The companion of this book is Volume 1 of the

Treatise on Basic Philosophy

SENSE AND REFERENCE

TABLE OF CONTENTS